Style and Ethics of Communication in Science and Engineering

Style and Ethics of Communication in Science and Engineering
Jay D. Humphrey and Jeffrey W. Holmes

ISBN: 978-3-031-79320-2 paperback

ISBN: 978-3-031-79321-9 ebook

DOI: 10.1007/978-3-031-79321-9

A Publication in the Springer series

SYNTHESIS LECTURES ON ENGINEERING #9

Lecture #9

Series ISSN

ISSN 1939-5221   print
ISSN 1939-523X   electronic

Style and Ethics of Communication in Science and Engineering

Jay D. Humphrey
Texas A&M University

Jeffrey W. Holmes
University of Virginia

SYNTHESIS LECTURES ON ENGINEERING #9

ABSTRACT

Scientists and engineers seek to discover and disseminate knowledge so that it can be used to improve the human condition. *Style and Ethics of Communication in Science and Engineering* serves as a valuable aid in this pursuit—it can be used as a textbook for undergraduate or graduate courses on technical communication and ethics, a reference book for senior design courses, or a handbook for young investigators and beginning faculty members. In addition to presenting methods for writing clearly and concisely and improving oral presentations, this compact book provides practical guidelines for preparing theses, dissertations, journal papers for publication, and proposals for research funding. Issues of authorship, peer review, plagiarism, recordkeeping, and copyright are addressed in detail, and case studies of research misconduct are presented to highlight the need for proactive attention to scientific integrity. Ample exercises cause the reader to stop and think. *Style and Ethics of Communication in Science and Engineering* thus motivates the reader to develop an effective, individual style of communication and a personal commitment to integrity, each of which are essential to success in the workplace.

KEYWORDS

journal publication, theses, grant writing, peer review, oral presentations, authorship, record keeping, research misconduct

Preface

How to write well.
How to publish your results.
How to secure funding.
How to speak well.
How to ensure integrity.

This book was written to help address these important aspects of beginning a career in science and engineering. In essence, scientists and engineers seek to discover and disseminate knowledge so that it can be used to improve the human condition. Effective communication thus plays an essential role in promoting technical advances. Simply put, communication is the ability to express oneself in a way that is understood readily and clearly. There will be no impact of scientific or engineering discoveries without effective written and oral communication.

In sections on writing well, we focus primarily on style — that is, rules of usage as well as principles of composition and form — and draw heavily from Strunk and White (1979), Berry (1971), and Brogan (1973). Indeed, many illustrative phrases and sentences were inspired by or modified from these works. We thus note here our indebtedness to these outstanding works and the examples therein. We encourage the reader to consult these excellent resources as well.

Although written communication, particularly the archival journal article, is most important to the widespread and long-term advancement of science and engineering, oral communication plays a vital role. From didactic lectures by an instructor to entertaining presentations for a lay audience, oral communication has the potential to capture the imagination and promote the advancement of science and its applications. Similar to theater, oral communication requires one to "tell a story" well, that is, to package information in a way that is assimilated quickly and retained. Technical advances in audiovisual capability can aid tremendously in stimulating and capturing an audience and thus should be integrated thoughtfully within the oral presentation.

It takes a lifetime to establish a good reputation, but only a moment to destroy it. Integrity in the workplace is just as important as understanding well the basic principles of science and engineering or the basic operation of a scientific instrument. Yet, even within the narrow context of technical communication, it is impossible to articulate a prescriptive set of rules or procedures for acting ethically. Despite the increasing prevalence of courses in research ethics, surveys suggest that most students learn the ethics of research and communication primarily "on the job," principally from their research mentor. Good training in the ethics of research thus begins with selecting a mentor who values such training and seeks to develop integrity through regular discussion and introspection. One goal herein is to stimulate this process of interaction around major issues most likely to face scientists and engineers in documenting and reporting their research.

The overall goal of this short book is not to be a standalone source on matters of style (which is left to professors of English or communication) or ethics (which is left for professors of philosophy or law). Rather, it is meant to motivate the reader to develop an effective, individual style of communicating and a personal commitment to integrity simply because it matters. Hence, this book is written based on personal experiences of the authors in research and training in the biomedical sciences and engineering, including the development and delivery of related graduate courses at Texas A&M University, Columbia University, and University of Virginia. Nevertheless, one of the best ways to learn to write well is to read extensively the works of good writers; one of the best ways to learn to speak well is to listen carefully to good speakers; one of the best ways to ensure integrity in the workplace is to learn from reputable professionals. The reader is thus strongly encouraged in this regard and, indeed, to keep a notebook wherein personal experiences and helpful observations can be recorded and reviewed periodically. Best wishes.

Acknowledgments

We thank Jodi Eyhorn, from the Department of Communication of Texas A&M University, for expert assistance in correcting early portions of the manuscript. We also thank Joel Claypool, of Morgan & Claypool, for patience and support during the writing process.

Portions of this work began via a Special Opportunity Award from the Whitaker Foundation. Finally, J.D.H. thanks Carolyn S. and Tommie E. Lohman for their continued support of many different educational initiatives at Texas A&M University, including composition of portions of this work.

Contents

CHAPTER 1

Motivation

The first part of this book is about communicating well, which is just as important to success in the workplace in science and engineering as it is in professions such as business, law, politics, and theology. Although there are useful guidelines for communicating well, there are no unique formulas. Indeed, individual differences can bring a freshness and vitality to a field; individual personalities can generate excitement and interest. *Each person should thus develop a style that is most effective for him or her.* The second part of this book addresses the need for professional responsibility, that is, integrity in the workplace. It has been correctly said that it takes a lifetime to establish a good reputation, but only a moment to destroy it. There is, therefore, a need for consistent and concerted attention to ethical conduct and the appropriate communication of research findings. This, too, requires thoughtful, personal commitment — it is more than simply trying to follow the rules, for rules may change, it is doing the right things for the right reasons. Consequently, this book is different from most textbooks in science and engineering. We seek to cause one to stop, contemplate, and adopt a personal course of action. In some sections, therefore, the style is more like a workbook. Indeed, as a point of departure, let us consider the following.

Exercise 1.1   List five of the most important inventions of all time.

    1.

    2.

    3.

    4.

    5.

Exercise 1.2   List five of the most important scientists, mathematicians, or natural philosophers of all time.

    1.

    2.

    3.

    4.

    5.

If you are like many of our previous students, you listed among your selections of the most important inventions of all time the airplane, automobile, computer, electric motor, refrigeration, steam engine, telephone, and so forth. Each of these inventions is truly great, and so too many others, but did you consider the invention of the *printing press with movable type*? If not, you are not alone. Yet, take a few minutes and imagine what the world would be like without books or scientific periodicals. Indeed, think about how the development of science, medicine, and engineering may have differed over the past five centuries had the printing press not been invented. For this reason, *Time-Life* magazine selected the printing press as the most important invention of the second millennium.

The rapid growth of printing with movable type reveals its overall importance. Invention of movable type is generally attributed to Johann Gutenberg (ca. 1397–1468), and the first book so printed is the famous Gutenberg Bible, which was completed in 1454/1455 at Mainz. By 1480, 111 towns throughout Europe boasted printing presses, and by 1500, this number grew to more than 238 (Boorstin, 1983, p. 270). In addition to the printing of the Bible, which had a significant influence on the development of science and Western culture (Dillenberger, 1961; van Doren, 1991), these presses allowed the printing and widespread distribution of classics by Aristotle, Cicero, Euclid, Plutarch, Ptolemy, and many others. It is thus thought that Gutenberg's invention played a singularly important role in the European Renaissance.

Recalling Exercise 1.2, many scientists, mathematicians, and philosophers deserve recognition as great. Among them, you may have listed Socrates, Plato, Aristotle, Archimedes, Copernicus, Galileo, Newton, Euler, Lavoisier, Gauss, Darwin, Maxwell, Planck, Einstein, or Pauling. How is it that we know about these great investigators? How is it that we know what they accomplished? Consider Sir Isaac Newton (1642–1727), for example, who is universally listed as one of the greatest natural philosophers of all time. Many know of Newton based on comments in courses on physics related to his law of gravitation, his laws of motion, or perhaps his experiment with a glass prism that revealed a spectrum of colors in sunlight. Fewer people know about Newton through in depth study, for example, by reading books such as *The Life of Isaac Newton* by Westfall (1993). Still fewer yet know of Newton because they have read his great works, his *Principia* of 1687 or his *Opticks* of 1704. Regardless of the particular path, we all know of Newton primarily through the written word, not oral tradition and certainly not first-hand interaction.

When reading about Newton, it is interesting to learn that he abhorred criticism of any kind and, in particular, interpersonal conflicts. Indeed, it appears that he was so concerned about criticism, especially from R. Hooke (1635–1703), then secretary of the Royal Society of London, that for many years he had little interest in publishing his greatest ideas. Apparently, the *Principia* was published only because of the persistent encouragement and personal financial sacrifice by Edmund Halley (1656–1742). This is remarkable! Similarly, it seems that Newton withheld publication of his *Opticks* until just after the death of Hooke. What if Newton had died first? Can we imagine how the development of science may have differed had Newton not revealed his brilliant thoughts through these two books?

Likewise, it is interesting to contemplate the development of Western society, which depended so strongly on Greek thought, had it not been for Plato (ca. 427–347 BC). It seems that Plato's mentor, the great Socrates, was content to lecture or discuss rather than to write or dictate. Although it is not clear how much of Plato's writings truly reflect Socrates, the importance of works such as *The Republic* is without question. The written word and its widespread distribution has impacted the world like few other things — it is fundamental to communication among scientists and engineers as well as the general public.

> Communication is defined as follows: To make known; impart. To transmit; have an interchange, as of ideas. To express oneself in such a way that one is readily and clearly understood.

> *American Heritage Dictionary*

Whether to have a long-lasting impact on human history or simply to contribute to success in the workplace, communicating well is a vitally important skill for the scientist or engineer. Indeed, not only must one communicate well with colleagues or a technical boss, there is often a need to communicate with diverse scientists, engineers, clinicians, venture capitalists, or the general public. For example, the National Institutes of Health (NIH) is currently promoting the importance of "team science" in biomedical research, which depends strongly on effective communication among individuals having diverse backgrounds, and many universities in the United States are promoting the importance of translational research, which requires interactions among clinicians, scientists, engineers, and those in business. Hence, we cannot overemphasize the importance of effective written and oral communication in research and development.

Because our focus is on technical communication, note that March 6, 1665, marks a beginning of the scientific periodical, based on the proceedings of the Royal Society of London entitled *Philosophical Transactions*. In the preface to the first issue, Henry Oldenburg (ca. 1617–1677) wrote (see Boorstin, 1983, p. 393):

> Whereas there is nothing more necessary for promoting the improvement of philosophical Matters, than the communicating of such, . . . ; It is therefore thought fit to employ the *Press*, as the most proper way to gratifie those, whose engagement in such Studies, and delight in the advancement of Learning and profitable Discoveries, doth entitle them to the knowledge of what this Kingdom, or other parts of the World, so,

from time to time afford. . . . All for the Glory of God, the Honor and Advantage of these Kingdoms, and the Universal Good of Mankind.

A careful editor would likely revise this Preface in the interest of conciseness and clarity, yet the message would remain apparent and the motivation to improve our communication skills certain. Indeed, from its beginning, the Royal Society of London sought clear communication in both written proceedings and oral presentations, not an "Artifice of Words." We are well advised to pursue the same today.

In conclusion, recall from the Preface that a good way to learn to write well is to read widely. The student having an interest in engineering, mathematics, medicine, philosophy, or science can learn much from the many books on the historical development of these fields. See Shamos (1959), for example, who provides brief background information on great physicists from Galileo to Einstein and includes excerpts from their original publications. Lightman (2005) provides a similar resource for scientists of the 20th century, and Clendening (1960) reprints portions of the early great publications in medicine. Other books of interest include Bell (1986), Mason (1962), Motz and Weaver (1989), Tarnas (1991), and van Doren (1991). It is interesting to conclude, consistent with aforementioned comments by Boorstin (1983), that Shamos (1959) observed: "The exchange of knowledge, facilitated by the publication of scientific journals, became — and remains — one of the most significant factors in the growth of physical science."

Exercise 1.3   Write and submit a three-page (double-spaced, 1-inch margins, 12-point font) essay on the role of printed books and scientific periodicals during the Age of Enlightenment.

Exercise 1.4   Write and submit a three-page (double-spaced, 1-inch margins, 12-point font) essay on the development of the Royal Society of London and its role in the advancement of science.

Exercise 1.5   Write and submit a three-page (double-spaced, 1-inch margins, 12-point font) report on the origins of the university and how it differed from the scientific academies of the 17th and 18th centuries.

CHAPTER 2

Writing Well

Vigorous writing is concise. A sentence should contain no unnecessary words, a paragraph no unnecessary sentences, for the same reason that a drawing should have no unnecessary lines and a machine no unnecessary parts. This requires not that the writer make all his sentences short, or that he avoid all detail and treat his subject only in outline, but that every word tell.

W. Strunk Jr. and E.B. White, (1979, p. xiv)

2.1　OVERALL APPROACH

Two of the most difficult aspects of writing are "getting started" and "finishing well." In other words, once we truly get started, it is usually easy to continue our line of thinking and to produce a first draft. Revising and polishing the first draft often takes longer than the initial writing, yet this is time well spent. Consider, therefore, some general guidelines for writing well, including a simple five-step recipe for completing a technical document:

1. Outline in detail.
2. Write freely.
3. Edit critically.
4. Read out loud.
5. Have a colleague proofread.

Although these steps may seem obvious, even trivial, they serve as important reminders and aid greatly in the composition of each new work no matter the level of one's experience.

2.1.1　Outline

Most writers agree that it is useful to begin with a detailed outline. Such an outline should contain the major headings that guide the flow of the work, but perhaps more importantly, it should also contain subheadings with bullets that highlight and order the major points within each section.[1] Each document is different, thus we should not feel compelled to force our presentation to fit within the confines of a particular outline. Nonetheless, most technical works adhere to the following basic outlines:

The technical proposal

- Project summary
- Specific aims
- Background and significance
- Preliminary results
- Research plan
- References

The technical paper or report

- Abstract
- Introduction
- Methods
- Results
- Discussion
- References

The M.S. thesis or Ph.D. dissertation[2]

- Abstract
- Introduction
- Background
- Methods
- Results
- Discussion
- Conclusions and recommendations

---

[1]Some encourage full sentences rather than bullets to document the main ideas in each subsection or paragraph, sentences that may later be used directly in the document. This, however, is a matter of personal style.

[2]Many European dissertations are very different. They consist of a sequence of individual chapters, each similar to a technical paper, all of which are tied together via short introductory and concluding chapters.

- References
- Appendices

Because of their importance, we discuss these different forms of technical communication in detail in Chapters 3 and 4. Here, we simply emphasize that a basic outline is the first step toward successful writing; it organizes the flow of the presentation and reminds us to address particularly important issues. Additional bulleting within each subsection further directs the writing. Indeed, with the use of word processors, one may easily use the final outline as a beginning document.

In summary, as in any activity, we increase our chances of "reaching the destination" when we have a map or detailed instructions to show the way. Note, therefore, that an outline will tend to be most useful and focused when it is constructed against the background of two questions (Gibaldi, 1995):

What is the overall goal that you wish to achieve with the document?
Who is the intended audience?

Toward this end, it is useful to critique the final outline with regard to both consistency and conciseness. For example, do points made in the introduction set up well the key points made in the discussion? Moreover, we tend to allow space in our outline for all of the information that we have collected during our research; we should, however, delete information that unnecessarily duplicates other documents or simply is irrelevant or unnecessary. Once done, it is then time to begin the actual writing.

2.1.2 Write Freely

One of the biggest impediments to writing efficiently and effectively is untimely self-criticism. How many of us have labored over that first sentence or first paragraph, rewriting and editing to the point of fatigue or frustration? Such editing is essential, but it is productive only if addressed at the right time and in the right way. By writing freely, we mean the unencumbered recording of a logical thought process. Indeed, it is often useful to disable the spell- and grammar-checking capabilities of word processors during the initial writing, for they contribute to the distractions of worrying about the initial spelling of words, ordering of phrases, and even punctuation. These and similar issues are addressed easily once we complete the initial draft. Indeed, we likewise should not initially worry about emphasizing active voice, ensuring sufficient variety in our word choice, focusing on conciseness, and so forth. Rather, at this early stage of composition, the most important thing is to get the major ideas onto paper (or the screen) and organized roughly in the right order.

2.1.3 Edit Critically

Once the first draft is finished, it is usually best (if time permits) to put it aside for at least a few days before beginning to edit critically. The reason for this is that we often see "what should be there" rather than "what is there" when we proofread our documents. Most of the remainder of this chapter addresses specific aspects of editing critically, which typically includes adding, deleting, and rearranging text.

The fundamental components of any technical document are sentences and paragraphs. A *sentence* is a grammatical unit typically consisting of a subject and a predicate (which tells something about the subject). A simple example is — I am. R. Descartes (1596–1650) expanded this to read — I think, therefore I am. Clearly, important sentences need not be complex. A *paragraph* is a grammatical unit typically consisting of multiple sentences that together express a complete thought. Many suggest that the lead sentence of each paragraph should introduce the main idea of that paragraph and the final sentence of each paragraph should summarize the main thought. This simple guideline helps to minimize unnecessary generality, that is, it helps to keep the writer focused.

A stepwise approach to editing critically exploits these two fundamental units of composition. For example, many suggest that the first step should consist of reading the first and last paragraphs of the document to ensure a consistent introduction and conclusion. The second step should consist of reading the first sentence of each successive paragraph to ensure that the work flows logically. Indeed, some go further to suggest that one should be able to glean the salient points of a document by reading only the first sentence of each paragraph. Although we do not wish to suggest such a dogmatic approach, casual guidance can certainly come from such an exercise. The third step of critical editing is a careful evaluation sentence by sentence. In other words, while reading each sentence within context, we should ask if it is necessary, if it is consistent in tense, and if it as concise and clear as possible. This brings us to the fourth and last step of critical editing, an evaluation word by word. We should ask, for example, if we have avoided the use of jargon as well as redundant or unnecessary words and if the intended meaning of each word actually reflects its definition. Word choice is critical. From a pragmatic perspective, we can simultaneously evaluate sentence by sentence and word by word.

As noted previously, the importance of critical editing cannot be overemphasized, hence we return to this issue in detail in Sections 2.2 to 2.8. Here, however, let us finish our discussion of an overall approach to writing well. After we have outlined our work, written freely, and edited carefully, our next step should be to read the document out loud.

2.1.4 Read Out Loud

Although this step may seem trivial or perhaps uncomfortable, it is amazing how sensitive the ear is to effective writing — different tenses, logical sequencing, unintentional rhymes, the overuse of certain words, and so forth. We strongly recommend, therefore, that one read the document out loud before going to the final step, asking a colleague to provide constructive criticism.

2.1.5 Have a Colleague Proofread

Technical advances in science and engineering have been spectacular, and continued promises of important discoveries make these professions intellectually attractive. Because of the trend toward multidisciplinary teams, one of the most enjoyable aspects of these professions is the opportunity to work with colleagues from many allied fields. Consequently, there is not only a need to write concisely, but also to write clearly. Although it is common to have colleagues from these allied fields coauthor many of our works, it is essential to have others proofread our work. That is, even though we may know best what needs to be said, the definition of an effective paper, proposal, report, or book is one that is understood and valued by others — this is the goal of effective communication.

Classmates and colleagues tend to be busy, thus we sometimes hesitate to ask them to proofread our work. Yet, they too would appreciate having someone provide feedback on their work and consequently will many times agree to do so for you. Consider establishing reciprocal agreements, whereby you exchange documents to be proofread. This will not only help the author by providing specific feedback, chances are it will help the reader both directly and indirectly. One not only learns by reading, oftentimes going through a document carefully and looking for ways to improve its clarity and conciseness teaches us much more. This is similar to the adage, "the best way to learn something is to teach it to others."

If you ask someone to proofread your work, make sure to tell them that you want them to be "brutally honest" rather than overly concerned about being critical. Moreover, once you receive the feedback, be careful to avoid the two most common responses: either to ignore the comments because you "know" that you were correct in the first place or to incorporate the suggested changes without questioning. These two responses are equally inappropriate. If a colleague questions the way something is stated, particularly if based on deep knowledge of the technical area, this at least suggests that the text could be written more clearly. In other words, if they do not understand what you are trying to say, chances are others will likewise not understand. Consider revising the text along the lines they suggest or in another way; the important thing is to give the suggested revision careful consideration. Conversely, incorporating suggested changes without questioning is dangerous. The primary goal is to communicate most effectively that which you are trying to say. If your colleague's suggestion does that, great; if not, work to improve clarity and conciseness and perhaps have your colleague read it again. Often, it is helpful to ask them what was confusing or what they thought you meant to say. Sometimes an explanation reveals how best to say it in the written word.

Finally, remember that it is good to keep marked manuscripts to evaluate them for possible consistent errors or patterns. In this way, we can become proactive in avoiding problems, whether it is an overuse of passive voice or an inappropriate use of modifiers. Being conscious of potential errors is the first step to avoiding them.

Exercise 2.1   Recover from your files at least three documents you have written and that were evaluated/graded on writing style. Compile a list of errors that occurred repeatedly and write brief examples of how to correct these problems.

Exercise 2.2   Read a journal paper in your area of expertise and record at least 10 sentences that could be improved for conciseness or clarity. Next, reread the paper out loud and record another five sentences that you could improve; note the types of concerns that are identified more easily when heard. Finally, suggest possible improvements for each of the 15 sentences.

2.2   REMOVING REDUNDANCIES AND UNNECESSARY WORDS

Now that we have a feel for an overall approach to writing well, let us begin to address specific aspects of "critical editing." Recall that effective technical writing is first and foremost clear and concise, which for obvious reasons is better written "Recall that effective technical writing is clear and concise." One way to ensure such characteristics in our writing is to *remove redundancies and unnecessary words*, sentence by sentence. Let us consider a few specific examples below (note: the original version is on the left and the corrected version is on the right, hence it is best to cover the right side first and consider how you might improve each example before looking at the suggested change):

The cells were cultured for a period of three weeks.	The cells were cultured for three weeks.
The temperature of the chamber remained between 35 and 39°C.	The chamber remained between 35 and 39°C.
The associated mechanisms are not known at this time.	The associated mechanisms are not known. (or, . . .remain unknown.)
The experiments were performed over a period of 10 hours.	The experiments were performed for 10 hours.
The new transducer is much smaller in size, which simplifies the design.	The new transducer is smaller, which simplifies the design.
The temperature increased at a rate of 3°/min.	The temperature increased at 3°/min.
The signal is lost below a threshold level of 10 Hz.	The signal is lost below a threshold of 10 Hz.

This thesis reports work done during the period from January 1998 to December 2000.	This thesis reports work accomplished from January 1998 to December 2000.
The algorithm searches outward from the center location.	The algorithm searches outward from the center.
The A/D converter allows a maximum of eight input signals.	The A/D converter allows eight input signals.
The range of the output signal was from a minimum of 2 to a maximum of 5 volts.	The output signal ranged from 2 to 5 volts.
The results of our experiments support the established theory.	Our experiments support the established theory.
Turn the potentiometer in the clockwise direction to increase the gain.	Turn the potentiometer clockwise to increase the gain.
Use the lenses that are convex in shape.	Use lenses that are convex.
The problem should first be formulated and then solved.	The problem should be formulated, then solved.
The amount of noise will be excessive if the signal is not filtered.	The noise will be excessive if the signal is not filtered.
There is no known analytical solution to this equation at this time.	No analytical solution exists for this equation.
The biopsy should be redesigned in the future to minimize the amount of tissue needed.	The biopsy should be redesigned to minimize the tissue needed.
The reason for this difference can be attributed to. . .	This difference can be attributed to. . .
Remember to remove the specimen during the calibration procedure.	Remember to remove the specimen during calibration (or, when calibrating).
There is a growing body of evidence that the hypothesis is indeed true.	There is growing evidence that the hypothesis is true.
After one test, there should be a sufficient quantity of culture media for a second test.	After one test, there should be sufficient culture media for a second test.

In the first example, the use of the word *weeks* implies a period or duration, which is therefore not needed. In the second example, the use of the unit °C specifies that the numerical value refers to a temperature, which thereby becomes redundant. Similarly, in the third example, we see that if something is not known, it is implied that it is not known at this time, which is thereby unnecessary. In hindsight, the other examples are likewise clear. Indeed, because these specific examples highlight a redundancy or unnecessary words, they may seem so obvious that we would be surprised if we ever wrote such sentences. Upon close examination of our previous works, however, we often find similar or even more flagrant examples. It is for this reason that we must be conditioned to look for redundancies and unnecessary words, which is often best learned via examples; see Brogan (1973) for additional examples.

Exercise 2.3   Find a technical research paper that you have written and scan it specifically for examples of redundancies or unnecessary words. Record five examples below with possible revisions.

If we read a number of published technical papers for style, we quickly realize that we could make many commonly used phrases more concise or even omit them without a loss of clarity. For example, how many times have we read the phrase "The purpose of this paper is to present. . . ," which we could write more concisely as "This paper presents. . . ." Consider the following phrases (left side) that occur frequently but can often be rendered better or omitted (right side) as follows (cf. Brogan, 1973; Valiela, 2001):

is used to develop	develops
is dependent on	depends on
We propose to use the combination of	We propose to combine
results in the simplification of	simplifies
It is interesting to note that	Note that (or, omit)
due to the fact that	because
in order to	to
in spite of the fact that	despite
as a result of	Omit
appears to be	seems
experienced a peak at	peaked at
in the event that	if
was found to be	was
a number of	many (or, various)
may be a mechanism responsible for	may cause
It is well-known that	Omit
for a long period of time	for a long period
is described in detail in	is detailed in
in the absence of	without
It is not uncommon that	It is common that
The finding is not inconsistent with	The finding is consistent with

Note, in particular, that the last two entries in this table emphasize that double negatives should be avoided. Moreover, see Appendix 2 in Day and Gastel (2006) for an expanded list of words and expressions to avoid.

In addition to removing redundancies and unnecessary words, there are many opportunities to *introduce conciseness* via word choice and sentence structure. For example, consider the following examples:

Because the structure is assumed to remain circular, . . .	Assuming the structure remains circular, . . .
This will enable us to develop a better understanding of. . .	This will enable us to understand better. . .
This finding is the opposite of that reported by. . .	This finding is opposite that reported by. . .
The model is capable of describing. . .	The model can describe. . .
Table 1 is a list of all findings. . .	Table 1 lists all findings. . .
The next section is a brief description of the experimental methods.	The next section briefly describes the experimental methods.
The faculty advisor was the supervisor of both the undergraduate and the graduate students.	The faculty advisor supervised both the undergraduate and the graduate students.
The temperature readings will be dependent upon the contact stress. . .	The temperature readings will depend upon the contact stress. . .
Our laboratory technician also serves as the budget manager.	Our laboratory technician also manages the budgets.
The following example is an illustration of the basic concepts of. . .	The following example illustrates the basic concepts of. . .

Before continuing, note that some of the suggested changes in the right-hand column assume a particular style not accepted by all technical writers. Some suggest that a table cannot list, a figure cannot show, a model cannot describe, a paper cannot present, and so forth. That is, some argue that only people can list, show, describe, or present; results are simply listed in a table, shown in a figure, described by a model, or presented in a paper. Because this is a matter of style, one must decide what approach to take, then be consistent.

As we shall see in Chapters 3 and 4, removing redundancies and unnecessary words not only results in writing that is clearer and more concise, it enables us to meet stringent limitations on words or pages in published works or proposals. Consider two instructive exercises:

Exercise 2.4 Write a three-page (double-spaced, 1-inch margins, 12-point font) biography of a leading scientist. Work hard to write clearly and concisely. Two days after finishing your essay, edit it further to reduce it to a single page without losing significant content. You will be surprised how easy and yet how powerful this exercise is. Finally, note that a one-page "white paper" is often all that is used to render important decisions in many professions; thus, it is important be able to write an effective short report.

Exercise 2.5 Use the "Word Count" tool in your word processor to determine the number of words in a short document (e.g., abstract) that you recently composed. Once done, set out to reduce the length of the document by 50% without compromising the content.

2.3   ACTIVE VOICE, FIRST PERSON, AND DIFFERENT TENSES

2.3.1   Voice

In *active voice*, the subject of the sentence performs the action indicated by the verb. Conversely, in *passive voice*, the subject of the sentence receives the action of the verb. The simple example below distinguishes between passive and active voice:

> Passive voice: The data were analyzed by him using an ANOVA.[3]
> Active voice: He analyzed the data using an ANOVA.

Although passive voice is acceptable, indeed sometimes more appropriate, most writers agree to prefer active voice, for it engenders conciseness and directness. In the example given, we see that seven words suffice rather than nine — this reduction represents a savings of approximately 20%. Given a 10-page paper, a 20% reduction in the number of words would yield an eight-page paper or else would provide an extra two pages to include more information; such savings can be significant. Moreover, comparing the two sentences in this example reveals the increased directness of the active voice, which promotes clarity and conciseness.

Albeit preferred, active voice is less common than passive voice in scientific writing. A simple change in the preceding example illustrates one reason for this:

---

[3] ANOVA is a common acronym for analysis of variance in statistics.

Passive voice: The data were analyzed using an ANOVA.
Active voice: We analyzed the data using an ANOVA.

In this case, each sentence contains seven words, thus the active voice does not increase conciseness. Moreover, the context can imply the "we" in the case of the passive voice, thus there need not be a difference in clarity (e.g., who did what). In many cases, authors prefer not to write in the first or the third person and revert to passive voice. The issue of person is addressed in the next section (or should we say, we address the issue of person in the next section). Here, however, consider examples of passive voice (left) and easy ways of changing them to active voice (right). First, changes primarily in verb form can be effective:

The specimen is connected to the device through a custom cannula.	The specimen connects to the device through a custom cannula.
The output signal is fed into a signal conditioner.	The output signal feeds into a signal conditioner.
In the next section, the underlying theory is given.	The next section gives the underlying theory.
In our current research, attention is directed to finding the mechanism.	Our current research directs attention to finding the mechanism.
The theory is dependent on five basic postulates.	The theory depends on five basic postulates.
X was used to create a surface-confined computational mesh.	X created a surface-confined computational mesh.
Increasing evidence has implicated the importance of. . .	Increasing evidence implicates the importance of. . .
Three different sectioning planes were used to form. . .	Three different sectioning planes formed. . .
Experimental noise is increased when unshielded cables are used.	Experimental noise increases with the use of unshielded cables.
A reader's attention is increased by the liberal use of figures and schematic drawings.	A reader's attention increases with the liberal use of figures and schematic drawings.

Second, changing the subject, which often necessitates changing the order of the words in the sentence, is often equally effective:

The specimen is connected to the device through a custom cannula.	A custom cannula connects the specimen to the device.
The results of the study are listed in Table 1.	Table 1 lists the results of the study.
The control is simplified by using commercial software.	Commercial software simplifies the control.
An improved result is obtained by refining the computational grid.	A refined computational grid improves the result.
The proper use of the equipment is described in Chapter 2 of the manual.	Chapter 2 of the manual describes the proper use of the equipment.
Ensure that all specimens are tested under the same conditions.	Use the same conditions to test all specimens.
The temperature is measured by a thermocouple.	A thermocouple measures the temperature.
These empirical findings are used as inputs into the theoretical model.	The theoretical model uses the empirical findings as inputs.
A detailed derivation of this equation is given in the appendix.	The appendix details the derivation of this equation.
The culture system is optimized by maintaining body temperature.	Maintaining body temperature optimizes the culture system.

A common excuse for (over)using passive voice is that it is natural because we often discuss past events, as, for example, "it was reported that" or "it was found that." Tense need not be the deciding factor, however, as revealed by the following simple example:

The pressure was measured by a mercury manometer.	A mercury manometer measured the pressure.
The pressure is measured by a mercury manometer.	A mercury manometer measures the pressure.

Finally, as noted previously, writing in the first or third person often allows us to avoid passive voice. Whereas the next section discusses the issue of person, as appropriate in scientific writing, consider the following, which revisit previous examples with alternate changes:

The specimen is connected to the device through a custom cannula.	We connected the specimen to the device using a custom cannula.
The results of the study are listed in Table 1.	We list the results of the study in Table 1.
These empirical findings are used as inputs into the theoretical model.	We used the empirical findings as inputs into our theoretical model.
The temperature is measured by a thermocouple.	We measured the temperature using a thermocouple.
A detailed derivation of this equation is given in the Appendix.	I derive this equation in detail in the Appendix.
Experiments were performed in triplicate for each set of...	We performed three experiments for each set of...
The culture system is optimized by maintaining body temperature.	We optimized the culture system by maintaining body temperature.

In summary, we do not have to avoid passive voice at all costs; indeed, passive voice is preferred in many cases. We also do not need to invoke first person to avoid passive voice. Nevertheless, *our general guideline is to prefer active voice when editing critically.*

Exercise 2.6   Select a journal paper in your field that interests you and scan it specifically for examples of passive voice. Record five examples below with possible revisions.

2.3.2   Person

Students of the history of science know that scientific writing used to be much more personal. As a simple example, consider the following excerpt from one of the works of W. Harvey (1578–1657) on the motion of the heart, (Clendening, 1960, p. 159):

> Besides the motions already spoken of, we have still to consider those that appertain to the auricles. Casper Bauhin and John Riolan, most learned men and skillful anatomists, inform us from their observations, that if we carefully watch the movements of the heart in the vivisection of an animal, we shall perceive four motions distinct in time and in place, two of which are proper to the auricles, two to the ventricles. With all deference to authority I say, that there are four motions distinct in point of place, but not of time. . . .

If written today, we may well have read (with little other editing):

> Besides the motions already noted, there is a need to consider those concerning the auricles. Bauhin (16xx) and Riolan (16xx) report that careful monitoring of the heart in an open-chest animal reveals four motions distinct in time and place, two of the auricles and two of the ventricles. Nevertheless, it is suggested here that these four motions are distinct in place but not time. . . .

Why has scientific writing become so impersonal today? Certainly, there has been an appropriate move away from the verbose, from patronizing prose, and from self-aggrandizement. Nevertheless, science and engineering are personal — they are advanced by people, usually for the good of people — and it is not only acceptable, in many cases it is more honest, direct, and effective to write in first person. For example, in Chapter 4 on writing research proposals, we will see that an important part of the NIH-R01 grant is a section called Preliminary Results. Imagine that you review such a section and read, "It has recently been shown that . . . (12)." Noting that (12) denotes reference number 12 in the list of references, the reviewer would not know if it was the applicant or another investigator who showed this important finding unless he/she looked at the reference list. Conversely, there would be no ambiguity if the applicant wrote "We recently showed that . . . (12)." In the case of a research proposal, clearly demonstrating one's previous work may increase tremendously the chances of funding, thus employing first person may be both effective and advantageous.

As a reminder that first person can yield effective and memorable prose, recall the following sentences from the seminal paper by James Watson and Francis Crick on the structure of DNA:

> We wish to suggest a structure for the salt of deoxyribose nucleic acid (D.N.A.). . . . It has not escaped our notice that the specific pairing we have postulated immediately suggests a possible copying mechanism for the genetic material.

There are three persons in typical grammatical structure: *first person* refers to the person or persons who are speaking or writing; *second person* refers to the person or persons spoken or written to; and *third person* refers to person(s) spoken or written about. For example, consider the common pronouns, singular and plural, in the three persons and three common cases (Vivian and Jackson, 1961):

SINGULAR	FIRST PERSON	SECOND PERSON	THIRD PERSON
Nominative	*I*	*You*	*He, She, It*
Possessive	*My, Mine*	*Your, Yours*	*His, Hers, Its*
Objective	*Me*	*You*	*Him, Her, It*

PLURAL	FIRST PERSON	SECOND PERSON	THIRD PERSON
Nominative	*We*	*You*	*They*
Possessive	*Our, Ours*	*Your, Yours*	*Their, Theirs*
Objective	*Us*	*You*	*Them*

Whereas the words *he* or *him* were used generically in the past to denote males or females, modern writers tend to be much more sensitive to issues of gender. Thus, there has been a move to use neutral pronouns. For example, the famous imperative from Star Trek fame, "To boldly go where no man has gone before," can be written as "To boldly go where no one has gone before." It is also acceptable to write *he/she* or *him/her* when desired, but we should prefer neutral constructions.

Finally, numerous terms such as *department chairman* or *layman* can be written as *department chair* or *layperson* to avoid this issue.

Albeit largely a matter of style, we suggest that it is acceptable and many times preferable to use a personal style in scientific writing. For example, it is acceptable to write: "Although they were the first to exploit their novel empirical observations by identifying quantitative correlations, we were the first to develop a theoretical basis to explain the observations."

As food for thought, consider the following simple examples as you decide on a particular style:

The authors recommend, therefore, that. . .	We recommend, therefore, that
Hence, it is suggested that. . .	Hence, I suggest that. . .
It will be seen that. . .	You will see that. . .
Based on these results, it was decided that. . .	Based on these results, we decided that. . .
It has been shown previously that. . .	We previously showed that. . .

One important warning, however, is that when "I" is used, be careful not to give the impression that it serves an egotistical end.

2.3.3   Tense

Tense is a property of time; it signifies when events occur or when conditions exist (Vivian and Jackson, 1961). There are six tenses: *past, present, future, past perfect, present perfect,* and *future perfect.* The perfect tenses typically involve the use of the words "have" or "had." Consider the following simple examples:

Past: I completed the experiment.
Present: I am completing the experiment.
Future: I will complete the experiment.
Past perfect: I had completed the experiment.
Present perfect: I have completed the experiment.
Future perfect: I will have completed the experiment.

Two of the key questions in scientific writing are, "What tense should I use when reviewing what others reported previously?" and "What tense should I use when reporting what I did?" Although it is of little comfort, the answer to these questions is that there is no set answer.

Some authors suggest, however, that if concepts or findings reported in a previous peer-reviewed work remain true, one should refer to them in the present tense. As a simple example, consider Newton's second law of motion, which was put forth in the 17th century. One could write "As Newton showed in the *Principia*, force equals mass times acceleration." Alternatively, one could write "As Newton showed in the *Principia*, force equaled mass times acceleration." All should agree that if it is still believed that force equals mass times acceleration, then present tense should be used. A more modern example could be, "Smith et al. (1999) show that . . ." versus "Smith et al. (1999) showed that. . . ." Again, the choice is largely a matter of personal style; the most important thing is to be consistent within a given paper.

Most authors agree that we should use past tense when reporting our own new findings, for they have not yet been verified or accepted widely. Hence, when writing the results section of a paper, it is appropriate to use "we measured" and "we found" or similar constructs.

2.4   INFINITIVES AND MODIFIERS
2.4.1   Infinitive

An infinitive is a verb form, a characteristic sign of which is the word *to*, for example, "to measure," "to quantify," or "to report" (Vivian and Jackson, 1961). A split infinitive occurs when a word or phrase separates the "to" and its complement. A famous split infinitive in recent years comes from the aforementioned quote from Star Trek: "To boldly go where no man has gone before," which we could rewrite as "to go boldly where no man has gone before." The issue is how we wish to go, boldly or fearfully. Although it is best not to split infinitives, grammarians are now less dogmatic with regard to this rule. Indeed, a purposefully split infinitive may be preferred in some cases. For example, consider the phrase "to promote exercise vigorously" (Iverson et al., 1998). There could be confusion by some as to whether vigorously relates to promote or exercise, hence writing "to vigorously promote exercise" could be clearer, unless of course the intent was "to promote vigorous exercise." Strunk and White (1979) also note that the sentence "I cannot bring myself to really like the fellow" is clear, concise, and relaxed. Nevertheless, the general rule should be: *Do not split infinitives unless the sentence is less awkward when doing so.*

Let us consider a few examples of split infinitives and how to correct them easily.

The goal of this project is to better understand. . .	The goal of this project is to understand better. . .
We plan to quickly initiate the funded study.	We plan to initiate the funded study quickly.

It is difficult to separately control X and Y. . .	It is difficult to control X and Y separately.
. . . , they failed to correctly diagnose	. . . , they failed to diagnose
It is bad practice in the laboratory to arbitrarily stop an experiment.	It is bad practice in the laboratory to stop an experiment arbitrarily.
To effectively study the source of the error, . . .	To study the source of the error effectively, . . .
The sponsor requested us to, with all possible haste, complete the final report.	The sponsor requested us to complete the final report with all possible haste.

The last example in this table is a particularly flagrant abuse of the infinitive.

Other examples of split infinitives occur when a single "to" serves multiple infinitives. Whereas it is generally acceptable to write, "There is a need to assemble and test the device," rather than "There is a need to assemble and to test the device," it is also better to write "There is a need to assemble the device according the sponsor's specification, then to test it . . ." rather than "There is a need to assemble the device according the sponsor's specification, then test it. . . ."

Finally, note that infinitives can occur in active or passive voice and in past or present tense. In these cases, the infinitives may take different forms, such as:

Present active: to tell
Present passive: to be told
Past active: to have told
Past passive: to have been told

Hence, that a word or phrase appears between the "to" and its complement need not signal that an infinitive has been split.

2.4.2 Modifiers

Another mistake common in technical writing is the *use of nouns as modifiers*. A modifier is a word, phrase, or clause that renders another word or group of words more specific; two common kinds of modifiers are adjectives and adverbs. In contrast, a noun is a person, place, or thing. Perhaps it has been in the spirit of trying to write concisely that nouns have been misused frequently as modifiers. In a syndicated column, J.J. Kilpatrick noted a few examples from the *New York Times*: "their court victory," which is better written "their victory in court," and "close-knit classical music world," which is better written "close-knit world of classical music." Common examples in the technical literature include "material science" rather than "the science of materials" and "fluid mechanics" rather than "the mechanics of fluids." Yet, such constructions need not be considered problematic, which reminds us that certain cases are acceptable. More flagrant examples of noun modifiers exist in many scientific papers and should be minimized. Tabulated below are a few examples found in recently published works (which we do not cite so as not to criticize particular authors, for many, including us, are equally guilty):

The primary extracellular matrix components include. . .	The primary components of the extracellular matrix include. . .
When tissue temperature reached. . .	When the temperature of the tissue reached. . .
Force and length data were used to compute stresses.	Stresses were computed from data on forces and lengths.
An increased wall stiffness of the aorta. . .	An increased stiffness of the wall of the aorta. . .
Minimum residual microfibrillar function. . .	Minimum residual function of the microfibrils. . .
Ultrastructural analysis has begun to. . .	Analysis of ultrastructure has begun to. . .
. . .could detect molecular level changes. . .	. . .could detect changes at the molecular level. . .
. . .will use gene expression measurements to. . .	. . .will use measurements of gene expression to. . .
Changes in cell structure and function reveal. . .	Changes in the structure and function of cells reveal. . .
The resulting surface stress appears. . . .	The resulting stress at the surface appears. . .
. . .the ability of the cells to move into the wound area. . .	. . .the ability of the cells to move into the area of the wound. . .
. . .to undergo changes in contractile protein expression. . .	. . .to undergo changes in the expression of contractile proteins. . .
. . .organ development becomes highly sensitive to. . .	. . . .development of the organ becomes highly sensitive to. . .
. . .of the neonatal fibroblast. . .	. . .of the fibroblast in neonates to. . .

In contrast to previous tables of examples on redundancies, the "corrected" right-hand side here often resulted in a longer sentence or phrase. Again, it may have been in the interest of conciseness that nouns have come to be misused (left side). Nevertheless, one is well advised to use nouns properly.

Next, consider a few simple suggestions to promote the proper use of appropriate modifiers (adjectives and adverbs). Recall, that adjectives modify nouns, whereas adverbs can modify a verb, an adjective, or another adverb. Adverbs may come before, after, or between the words that they modify. When possible, a sequence of modifiers should be listed according to length or logical order. For example, Berry (1971) suggests that "tired, bored, and exhausted" is written better as "bored, tired, and exhausted" because it is likely that one becomes bored before tired. He likewise suggests that the modifiers "dry, withered, and flaky" should be ordered in the sequence in which they occur: "withered, dry, and flaky."

Finally, note that "a," "an," and "the" are called articles. The definite article "the" refers to something or someone in particular. Hence, when we read "A significant finding was…" versus "The significant finding was . . . ," we see that the former refers to one of many significant findings, whereas the latter refers to one finding that was significant. This simple distinction must be respected.

Although many modifiers are effective in different forms of writing, their overuse in scientific writing may suggest that one's results are not quantitative, that they need embellishing. For example, instead of saying that data "are very noisy," we need only say they "are noisy," then provide specific measures such as a signal-to-noise ratio to quantify the degree of the noise. Similarly, a numerical method may be "very robust," but if it is robust, that is all that needs to be said. In other words, the modifier "very" often adds very little (as in this case). Similarly, "quite" is quite unnecessary in most cases, including this one, and although the word "rather" is considered by some to be rather important, it often is not. A general rule, therefore, is: *Do not use modifiers unless the meaning is clarified by doing so.*

Exercise 2.7 Review a technical paper that you wrote previously and eschew all unnecessary or inappropriate modifiers. If you are somewhat puzzled why you used words such as "somewhat," take comfort that you are not alone. As examples, record five illustrative sentences below and possible corrections.

2.5   ADDITIONAL ISSUES OF WORD CHOICE

The best advice related to word choice is to keep a dictionary within easy reach and to consult it frequently. With regard to the five-step recipe for composition given in Section 2.1, however, we should remember that this should be done during the phase, "edit critically." Indeed, if you are struggling for just the right word while "writing freely," it is often best to put an "xx" in the text so that you are reminded to search for an appropriate word later and not interrupt the flow of your thoughts and composition. Here, however, we briefly identify and discuss some words that are often used interchangeably but should not be so used. Consider, for example:

> *Alternative/alternation:* An *alternative* is a choice between two mutually exclusive possibilities. An *alternation* is a successive change from one thing to another and back again.
>
> *Amount/number: Amount* refers to a quantity that is not countable, whereas *number* is used when it is possible to count. It is thus correct to say "The amount of information available was not sufficient for . . ." or "The data suggest a number of conclusions."
>
> *Because/since:* Strictly speaking, *because* refers to a cause–effect relationship and *since* refers to a past event. It is appropriate to write "Because the results suggested . . ." and "Since the last conference . . ."
>
> *Between/among:* In general, use the word *between* when considering two things and use the words *among* or *amongst* when dealing with more than two things.
>
> *Can/may:* The word *can* has to do with ability, whereas the word *may* has to do with having permission.
>
> *Compare with/compare to:* Use *compare with* when examining or discussing similarities or differences. In general, only use *compare to* when representing a metaphorical similarity.
>
> *Complement/compliment:* A *complement* is something that completes or brings to a whole. A *compliment* is an expression of congratulations or praise.

Comprise/compose: Comprise means to consist of or to include. *Compose* means to make up the constituent parts of, to constitute or form. Good examples are "The Union comprises 50 states" and "Fifty states compose (or constitute) the Union."

Continual/continuous: Continual means with occasional interruption, whereas *continuous* means without interruption.

Data/datum: Data are plural, typically representing facts or information. *Datum* is the singular form of data, often used in the context of a point from which to measure.

Due to/because of: Due to means attributable to. *Because of* relates to a cause or reason for occurring. A helpful hint is that a sentence should not start with *due to*.

Effect/affect: An *effect* is a noun; it implies a result, something that is caused. *Affect* is a verb; it brings about a change. To affect is thus to influence or impress.

Either/neither: It is correct to write "either A or B" and likewise "neither A nor B," but we do not use "neither A or B." Moreover, in each case, these words imply only two options, hence we cannot say "either A, B, or C."

Essential/important: Essential implies indispensable, fundamental, or absolute. *Important* merely implies significant or noteworthy.

Farther/further: Farther should be used when the context is distance. *Further* implies something in addition, such as the need for further experiments. Hence, one does not move a fixture further toward the center.

Good/well: In most cases, *good* is used to modify a noun (e.g., she is a good writer), whereas *well* is used to modify a verb (e.g., she writes well).

However/nevertheless: Strunk and White (1979) suggest that we should avoid beginning a sentence with the word *however* when the meaning is *nevertheless* or *yet*. This is easily corrected via replacement with these more acceptable beginning words or by moving the *however* to the middle or end of the sentence. When used at the beginning of a sentence, *however* should be thought of as "in whatever way" or "to whatever extent." A good example is given by Strunk and White, (1979, p. 49): "However discouraging the prospect, he never lost heart." In contrast, it would be better to write "Nevertheless, he never lost heart despite the discouraging prospects" rather than to write "However, he never lost heart despite the discouraging prospects."

Imply/infer: Imply means to suggest or indicate by logical necessity, whereas *infer* means to deduce based on available evidence.

Precede/proceed: Precede means to come before in time, to occur prior to. *Proceed* means to go forward, especially after an interruption, or to move on in an orderly fashion.

Principal/principle: A researcher may be the *principal investigator* on the project but not the *principle investigator*. One may use a scientific principle but not a scientific principal.

Another usage that is often confused is that the solution to an eigenvalue problem yields a principal value and in mechanics one may compute a principal stress or strain.

Shall/will: It is suggested by some that *shall* should be used for future expectations in first person and *will* should be used in second and third person. This distinction between *shall* and *will* occurs only in formal writing, however, and the word *will* often suffices. A good example is that *will* is appropriate in grant proposals, for example, "We will test the hypothesis that. . . ."

That/which: In general, use *that* to lead into a defining or essential clause and use *which* to lead into an inessential or nonrestrictive clause. Kilpatrick (1984) suggests an easy way to decide usage in most cases: use *which* whenever the clause is set apart by commas and use *that* otherwise. The key point, however, is the word *that* is used with essential clauses. For example, note the difference between the following sentences.

"The transducer that is broken is on the shelf."

"The transducer, which is broken, is on the self."

In the first case, only the transducer that is broken is on the shelf. In the second case, the transducer is on the shelf and it happens to be broken.

That/who: That refers to things and *who* refers to people.

While/whereas: Strictly speaking, *while* should be used to convey a sense of time, for example, "The computer acquired data while the device subjected the cells to increasing mechanical loads." Nonetheless, many accept *while* as a substitute for *although*. In contrast, *whereas* means "it being the fact that" or "inasmuch as."

Next, consider a few words that are useful in technical writing but sometimes misused.

Aforementioned: This word is an adjective; it must be combined with a previously used noun. For example, it is correct to write "The aforementioned finding suggests. . . ," but it is incorrect to write, "As aforementioned, ..."

And/or: This is a construction used by some, but often best avoided. Use either the word *and* or the word *or* as appropriate.

Correlate: To put into a complementary, parallel, or reciprocal relationship, not implying causality.

Dilemma: Either a situation that requires one to choose between two equally viable alternatives or a predicament that seemingly defies a satisfactory solution.

Former: The first mentioned of two things.

Latter: Like *former*, this word implies two choices. If one has a list of three or more items, then to refer to the last one in the list, simply say "the last one," not "the latter one."

Per: "Pursuant to"

This: For clarity, follow the word *this* with a noun. For example, do not write, "This is to be expected," but rather write, "This nonlinearity is to be expected" or "This finding is to be expected."

Finally, some words have specific meanings in science and mathematics even though they are often used loosely in everyday speech. Because our interest is scientific writing, however, we must respect the specific meanings. Three prime examples of such words are *significant*, *necessary*, and *sufficient*. It would be natural, for example, to write: "The response of Group A differed significantly from that of Group B." Yet, we must ask whether this is what we really mean. The word *significant* in science usually carries a statistical meaning, that is, it usually implies that based on a standard statistical test, there is a significant difference between two metrics (e.g., as indicated by a $p < 0.05$ associated with a specific statistical analysis). If such a test was performed and passed, then we could write our illustrative sentence as given; if not, it would be better to use a different word or to delete the modifier altogether. In the absence of a statistical test, it would be better to write: "The response of Group A differed markedly from that of Group B" or simply "The response of Group A differed from that of Group B." The words *necessary* and *sufficient* similarly have precise meaning in mathematics and they are often used together. In this context, *necessary* means required and *sufficient* means adequate. For example, a necessary and sufficient condition for a solution to hold is much stronger than a sufficient condition alone.

   In concluding this section, it is interesting to consider a comment ascribed to the famous ancient philosopher Socrates:

The wise man knows that he knows not; the fool knows not that he knows not.

Similarly, consider a comment by the famous modern philosopher Bertrand Russell:

Although this may seem a paradox, all exact science is dominated by the idea of approximation….When a man tells you he knows the exact truth about anything, you are safe in inferring that he is an inexact man.

If we accept that science represents relative, not absolute, truth, then should we be careful not to use strong phrases such as "the data demonstrate" or "the data prove." For example, should we use phrases such as "the data suggest" or "the data imply." Similarly, should we avoid saying that something "is," but instead say that "it appears that." Here, again, we simply suggest that one should think carefully about this issue, make a purposeful decision within the context of common usage, and be consistent.

Exercise 2.8   Read three technical papers in your field and generate a list of phrases that reflect either a "certainty" or a "possibility" with regard to important findings or conclusions. Write a one-page summary and indicate the approach to communicating such results that you find to be the best.

2.6   PUNCTUATION, ABBREVIATIONS, AND FOREIGN LANGUAGES

2.6.1   Exploit Methods of Punctuation

Punctuation is a system of devices or marks (e.g., commas, semicolons, colons, dashes, and parentheses) that clarify relationships between words and groups of words (Vivian and Jackson, 1961). Aside from the standard use of the period, many writers of science and engineering tend to use commas sparingly and to avoid using semicolons, colons, and dashes. Although we should not overuse such devices, variety in punctuation can be as effective in written communication as variety in tone can be in oral communication. We list here a few rules of punctuation, but we encourage the reader to give particularly careful thought to the effective use of semicolons, dashes, and parentheses. As a start, consider Rules 2 to 4 of Strunk and White (1979):

One should use a comma after each entry, except the last, in a list of three or more entries that share a common conjunction such as *and* or *or*. For example, we should write "this finding was unexpected, repeatable, and important." To appreciate this usage, recall from Section 2.1 that the fourth step in writing well is "read out loud." Doing so here, the ear reveals a difference between "this finding was unexpected, repeatable, and important" (with a verbal pause after each comma) and "this finding was unexpected, repeatable and important." In other words, the latter case sounds like the finding was "unexpected" as well as "repeatable and important."

When paired, commas are useful devices to set off a nonessential clause, for example, "The transducer, which is broken, is on the shelf." When used with a conjunction to introduce an independent clause, the comma should be omitted before the *and* when the clauses relate closely. In contrast, the comma should almost always precede conjunctions such as *but, for,* and *or*. Commas are also useful to set off an introductory phrase, such as "In this paper, we show. . . ." Finally, a comma can be used to separate three or more modifiers, such as in the case of a "randomized, double-blind, clinical trial."

Use the semicolon instead of a period when independent clauses relate closely and it is effective to highlight this similarity. The only exception to this rule is the case of short independent clauses. Consider, therefore, the following examples from Strunk and White (1979):

Stevenson's romances are entertaining. They are full of exciting adventures.	Stevenson's romances are entertaining; they are full of exciting adventures.

It is nearly half past five. We cannot reach town before dark.	It is nearly half past five; we cannot reach town before dark.
Man proposes, God disposes.	—
Here today, gone tomorrow.	—

The semicolon is also useful to separate main clauses that are joined by conjunctive adverbs such as the following: *indeed, yet, however, moreover,* or *hence*. For example, we might write (Iverson et al., 1998): "This consideration is important in any research; yet it is often overlooked, if not denied."

Use the colon before a long in-line quotation (see Section 2.7), to introduce a list, or to separate independent clauses when the first clause introduces the second one. For example, if we wish to specify the composition of a physiological solution used in an experiment, we might write the following.

The specimens were immersed in a physiological solution consisting of, in mM: 116.5 $NaCl$, 22.5 Na_2HCO_3, 1.2 NaH_2PO_4, 2.4 Na_2SO_4, 4.5 KCl, 1.2 $MgCl_2$, 2.5 $CaCl_2$, and 5.6 dextrose.

Like the comma, one can use short dashes (or em dashes) and parentheses to set off nonessential, but clarifying, clauses or entries. The decision to use the em dash or parenthesis (more common) is again a matter of style, with the em dash typically reserved for the longer, sometimes tangential, breaks in thought. Consider the following two cases:

Of the many risk factors for coronary artery disease — high cholesterol, high salt intake, cigarette smoking, lack of exercise, diabetes, and hypertension — some can be avoided by simple changes in lifestyle.

Many risk factors for coronary artery disease can be controlled by simple changes in lifestyle (e.g., cholesterol, high salt, and smoking).

Parenthetical setoffs are also useful in providing supplementary information or identifications. For example, it is common to read: "of the 10 tests, only 5 (50%) were successful," "the differences were not significant ($p > 0.05$)," or "a consistent volume of fluid (10 ml) was injected." As noted below, it is common to include clarifiers within parenthetical set offs such as (for example, . . .) or (that is, . . .), abbreviations for which are given below.

The hyphen has many uses as well; see Brogan (1973) for a good discussion of this device. Commonly misunderstood uses are numerical or multiword modifiers. For example, we should write "The diameter of the device is 5 mm," but we should use the hyphen to write "The 5-mm-diameter device. . . ." We should also use the hyphen to write out numbers such as *thirty-seven* or *two-thirds*. Another use of a hyphen is in the pairing of words that, via the natural evolution of grammar, often become single words. A simple example is *mechanical transduction*, which became *mechano-transduction* and now is *mechanotransduction*. Finally, multiword modifiers are often hyphenated, for example, "the signal-to-noise ratio" or "one-way Student *t*-test." Uses such as "pre- and post-surgical" are also common.

With regard to numbers, it is common to write out in words those numbers less than 10 (e.g., zero, one, two) but to write out numerically those numbers 10 or greater (e.g., 11, 100, 1000). There are exceptions, however (Blake and Bly, 1993). For example, if data are collected at days 0, 3, 7, and 14, we would not write "at days zero, three, seven, and 14." In other words, one of the best rules of thumb is consistency and clarity. Moreover, always write large numbers in a way that is most easily understood. For example, the number 30,000,000 may be best understood as 30 million if referring to dollars or population; in contrast, it may be best understood as 3×10^7 if referring to a quantity in physics or chemistry or 30 MPa if referring to stress in mechanics. The example of 30 MPa reminds us to use, when appropriate, accepted prefixes: giga (G), mega (M), kilo (k), milli (m), nano (n), and so forth. The best rule of thumb, therefore, is always write for clarity. Finally, it should be noted that decimal values less than unity should be written with a leading zero, for example, 0.15 rather than .15. Whether one uses decimal values or not, also remember to include only *significant digits*, that is, information that is reliable. For example, although a calculator or computer may provide an answer of 4.1248432, if only three of the digits are reliable then we should write this as 4.12. Refer to elementary textbooks on physics or chemistry for good discussions on the appropriate use of significant digits.

As last reminders, do not use the apostrophe in special cases of decades or centuries, rather one should write 1970s or 1700s. Words such as *its* and *it's* and *whose* and *who's* are often confused. This is simple to remember: *it's* and *who's* are contractions of "it is" and "who is," whereas *its* and *whose* show possession. Contractions should be avoided in formal writing, however. Finally, it is important to emphasize that most publishers *use a single space after a period, not two spaces*. Not only is the single space typically more pleasing to the eye, it is also an effective means to reduce the number of pages and hence cost of publication because those extra spaces add up.

2.6.2 Abbreviations

Many writers suggest that abbreviations should be avoided in formal writing. In technical writing, however, we should merely minimize the use of abbreviations, using them only when they improve conciseness or are common within the intended context. One of the easiest ways to decide whether to use a particular abbreviation is to ask if it will improve or impede the reader's understanding. For example, many readers of technical papers go directly to the figures or results to see what was found, or they go directly to the discussion to see what was deemed to be important. They can be frustrated, therefore, if the figure legends, results, or discussion contain uncommon abbreviations that require them to search the introduction or methods to find the associated meanings. This should be avoided. Nevertheless, many abbreviations are so common that it would be surprising if they were used with explanation. Examples include ANOVA (analysis of variance), DNA (deoxyribonucleic acid), ECG (electrocardiogram), MRI (magnetic resonance imaging), and mRNA (messenger ribonucleic acid). There are, of course, many similarly common abbreviations. Scientific units should also be abbreviated without definition, for example, kPa (kilopascal), MHz (megahertz), ml (milliliters) mmHg (millimeters of mercury), and mM (millimolar).

Many other abbreviations, such as LV (left ventricle) or MAP (mean arterial pressure), are used widely and so too for abbreviations of many biologically important molecules and chemical compounds. For example, one would be expected to use the following abbreviations: NO (nitric oxide), transforming growth factor (TGF), and (poly)methyl methacrylate (PMMA). In these cases, however, common practice is to introduce the abbreviation only if it is used three or more times in subsequent text and to define the abbreviation at its first occurrence in the body of the paper, not the abstract. It is best not to construct new abbreviations, however, just because a descriptor is used repeatedly. For example, we would not introduce NM for noun modifier even if used extensively. Similarly, we would not use SI for split infinitive. Indeed, this example reveals that one should be careful not to define new abbreviations that are identical to commonly accepted abbreviations (e.g., SI is French for *Systeme Internationale*, the common units of measurement in most of science and engineering). Iverson et al. (1998) provides an extensive list of accepted abbreviations in medicine.

2.6.3   Foreign Languages

Many publishers seek to reduce the length of a publication because additional pages translate into additional costs. For this reason, some well-accepted abbreviations are encouraged and thus are common. Four of the most common abbreviations come from Latin, namely:

1. *et al.*, which means "and others," is commonly used when referring to a publication by three or more authors. In such cases, it is customary to cite the last name of the first author followed by "and others," for example, Smith et al. (1999) or (Smith et al., 1999). Whether one italicizes the Latin *et al.* depends on the publisher, but in most cases, a period should follow the *al.*

2. *e.g.*, which means "for example," is often used in parenthetical situations (e.g., in this way). Remember, too, that an example is just that, one of many possible illustrations; it is not a unique clarifier.

3. *i.e.*, which means "that is," is also often used in parenthetical situations (i.e., it often appears in this type of context). In contrast to *e.g.*, using *i.e.* is similar to using the phrase "in other words" and thus is meant to clarify a meaning, not to provide an illustrative example.

4. *cf.*, which means "compare with," is often used to draw attention to a similar or related illustration, equation, or other scholarly work, for example, (cf. Figure 1).

Another common abbreviation from the Latin is,

5. *etc.*, which means "and other unspecified things of the same class" or simply "and so forth." Albeit commonly used, most grammarians agree that *etc.* should not be used in formal writing or, if so, only sparingly for good purpose. If one does not wish to provide an exhaustive list, using "for example" is an appropriate way to indicate the listing of some, but not all, of the members of that class. One would thus never use *e.g.* and *etc.* within the same parenthetical statement.

Although scientific and engineering documents should be scholarly, they should not be pretentious. An attempt to impress the reader by using phrases or words from Latin, Greek, or other "foreign" languages is not advised in general unless their meaning is well-known and they engender conciseness or clarity. For example, some words and phrases are common in the biomedical literature and should be used, for they are well understood. In addition to the aforementioned *et al.*, *i.e.*, *e.g.*, and *cf.*, consider for example:

de novo: anew
in situ: in its natural place
in vitro: "in glass" or generally in an artificial environment
in vivo: within a living organism
ex vivo: outside of a living organism, but still living

Other acceptable, but less common, examples are:

ad infinitum: without end or limit
in toto: totally, altogether, entirely
reductio ad absurdum: reduction to the absurd
status quo: as it is now

2.7   FOOTNOTES, QUOTATIONS, AND PROPER CITATION

2.7.1   Footnotes

Footnotes are brief notes placed at the bottom of a page that provide a citation (older use) or a comment on a specific part of the main text. Although scientists and engineers used footnotes extensively in the past, such usage is generally discouraged today. We do not advocate eliminating footnotes, but we do encourage sparse usage. For example, footnotes can provide brief examples or clarifications that do not otherwise fit within the text using parenthetical devices such as commas, parentheses, or em dashes. Footnotes should not be used to solve problems in organizing material or sentence structure, however.

2.7.2   Quotations

Quotations must be denoted in one of two ways: if integrated within the text, they must be enclosed within quotation (" ") marks; if longer, and singled out, they should be indented but appear without quotation marks. Some publishers also use a smaller font for indented quotations although we do not advocate this policy. For example, let us recall the quotation from W. Strunk Jr. that is given at the beginning of this chapter:

> Vigorous writing is concise. A sentence should contain no unnecessary words, a paragraph no unnecessary sentences, for the same reason that a drawing should have no unnecessary lines and a machine no unnecessary parts. This requires not that the writer make all his sentences short, or that he avoid all detail and treat his subject only in outline, but that every word tell.

In general, use longer block quotations sparingly, if at all, in a technical document. Many times, the reader will skip such quotations to get to the meaning or importance of the quotation that follows. Another rule of thumb is to ask whether the quotation is necessary or if it is simply an easier alternative. If the latter, a brief reference to the original source with original commentary would be better.

Ellipses, that is, three dots in sequence (. . .), indicate that words are omitted, usually from a quotation. Four such dots in sequence usually indicate that words are omitted at the end of a sentence, hence the last dot can be thought of as the period at the end of that sentence. Beginning a quotation with a lowercase letter indicates that the author has omitted the initial part of the quote; beginning with a capital letter indicates that one is beginning the quotation at the beginning of the sentence. Thus, ellipses are not needed at the beginning of a quotation.

When information is missing or incorrect in a quotation, it is acceptable to provide complete and accurate information. The information that is added should be enclosed by brackets [ ]. For example, given the quote "Newton postulated…," one may write "[Isaac] Newton postulated. . . ."

It is acceptable, however, to provide a quotation exactly as it appears without correcting obvious or subtle errors provided this does not mislead the reader. Finally, one may find or insert [sic] in a quotation. The form [sic] comes from the Latin and indicates that a seemingly paradoxical word, phrase, or fact is not a mistake; it should be read as given.

2.7.3   Proper Citation

Although we discuss issues of ethics in Chapters 6 to 8, including plagiarism, we briefly mention it here for convenience. Simply put, plagiarism is the passing off as one's own the ideas and words of another. Actually, "pass off" is too soft of a word; plagiarism is intellectual theft. Most universities have writing centers and associated Web pages, hence one can find formal definitions and excellent examples of plagiarism.

The best way to avoid plagiarism is through proper citation. Although we tend to learn in English classes that there is a particular way to cite works in a bibliography, in scientific writing, the citation format varies considerably from publisher to publisher. Hence, the best advice is to read the "instructions to authors," which can be found on the Web page for the journal or publisher or often within the journal itself. We give examples of different styles of citation in Section 3.1.9.

Exercise 2.9   "The Double Helix" by J.D. Watson is a wonderful account of the events that surrounded the discovery of the double-helix structure of DNA. Read this book and write a three-page summary that highlights issues of ethical interest.

2.8   VOCABULARY

Vigorous writing should be clear and concise; nevertheless, it should also be provocative and engaging. The reader is thus encouraged to read Chapter 5 in Strunk and White's *The Elements of Style*. There is a need to employ words of power (i.e., having strong meaning) without becoming verbose or haughty. One way to accomplish this is to expand our vocabulary, which is perhaps best accomplished by keeping a diary of words as we read technical papers and books. When we come upon a forceful, precise, or attractive word, we should take note of it. Knowing that the author may have misused the word, however, we should always consult a reliable dictionary when recording the associated definition. A good dictionary can be found online at www.m-w.com. Here, we list a few words that one can use advantageously in technical writing, which may or may not be used on a daily basis by the reader.[4]

---

[4]These definitions are taken largely from the American Heritage Dictionary.

Abate: To reduce in amount, degree, or intensity; lessen.

Adverse: Antagonistic in design or effect; hostile; opposed.

Ancillary: Subordinate.

Assume: To take for granted; suppose.

Attenuate: To make slender, fine, or small.

Augment: To make greater, as in size, extent, or quantity; enlarge.

Causal: Pertaining to or involving a cause.

Caveat: A warning or caution.

Cogent: Forcibly convincing.

Copious: Yielding or containing plenty; affording ample supply.

Corroborate: To strengthen or support; attest the truth or accuracy of.

Cull: To pick out from others; select.

Cursory: Hasty and superficial; not thorough.

Delve: To search deeply and laboriously.

Didactic: Intended to instruct; expository.

Disparate: Completely distinct or different in kind; entirely dissimilar.

Dubious: Fraught with uncertainty or doubt; uncertain.

Egregious: Outstandingly bad; blatant; outrageous.

Eminent: Towering above others; projecting; prominent.

Enigma: An obscure riddle; puzzling, ambiguous, or inexplicable.

Equivocal: Capable of two interpretations; cryptic; evasive; ambiguous.

Erudite: Deeply learned.

Exacerbate: To increase the severity of; aggravate.

Extant: Still in existence; not destroyed, lost, or extinct.

Extenuate: To lessen or attempt to lessen the magnitude or strength of.

Fortuitous: Happening by accident or chance; unplanned.

Fraught: Attended; accompanied.

Glean: To collect bit by bit.

Hypothesize: To assert a hypothesis (i.e., an assertion subject to proof).

Inadvertent: Accidental; unintentional.

Inchoate: In an initial or early stage; just beginning; incipient.

Inexplicable: Not possible to explain.

Inordinate: Exceeding reasonable limits; immoderate; unrestrained.

Integral: Essential for completion; necessary to the whole.

Intrinsic: Pertaining to the essential nature of a thing; inherent.

Lucid: Easily understood; clear.

Manifold: Of many kinds; varied; multiple.

Marked: Having a noticeable character or trait; distinctive; clearly defined.

Myriad: A vast number; a great multitude.

Nadir: The place or time of deepest depression; lowest point.

Nullify: To make ineffective or useless.

Obviate: To prevent or dispose of effectively; to render unnecessary.

Ostensible: Given or appearing as such; seeming; professed.

Ought: Indicates obligation or duty, prudence, or desirability. Use with *to*.

Paucity: Smallness of number; fewness.

Permeate: To spread or flow throughout; pervade.

Peruse: To read or examine, especially with great care.

Posit: To put forward as a fact or truth; to postulate.

Postulate: Something assumed without proof as being self-evident or generally accepted, especially when used as a basis for an argument.

Premise: A proposition upon which an argument is based or from which a conclusion is drawn.

Proliferate: To reproduce or produce new growth rapidly and repeatedly.

Promulgate: To make known by public declaration; announce officially.

Propensity: An innate inclination; tendency; bent.

Purview: The extent or range of function, power, or competence; scope.

Quiescent: Inactive or still; dormant.

Recant: A formal retraction of a previously held belief or statement.

Recondite: Not easily understood by the average person.

Reiterate: To say over again.

Replete: Plentifully supplied; abounding.

Requisite: Required; absolutely needed; essential.

Retrospect: A review, survey, or contemplation of things in the past.

Salient: Striking; conspicuous.

Spurious: Lacking authenticity or validity; counterfeit; false.

Substantiate: To support with proof or evidence; verify.

Succinct: Clearly expressed in few words; concise; terse.

Sundry: Various; several; miscellaneous.

Surmise: To infer without sufficiently conclusive evidence.

Tacit: Not spoken.

Tantamount: Equivalent in effect or value. Used with *to*.

Tractable: Easily managed or controlled; governable.

Ubiquitous: Seeming to be everywhere at the same time; omnipresent.

The space below allows you to record additional words, with their definitions, that you would like to add to your technical vocabulary.

2.9   CLOSURE

Recalling that one of the best ways to improve one's writing is to read widely, we should not only read for pure enjoyment or the gaining of new technical information, we should also read with the intent of learning how to write well. In other words, take note of the infinitives, the effective use of punctuation marks, and so forth; record and use particularly forceful words and phrases. Aside from scientific publications, works of history, philosophy, and theology (one of the four original academic disciplines, with law, medicine, and natural philosophy) often represent good examples of writing well.

As a specific example, consider the book *On Growth and Form* by D'Arcy Thompson (1917). In the foreword of the 1961 abridged edition, it is noted that P.B. Medawar wrote that Thompson's work was "beyond comparison the finest work of literature in all the annals of science that have been recorded in the English tongue." What gave rise to such a claim? Thompson was a true scholar, with expertise in the classics, mathematics, and zoology; moreover, he purposed both to document good science and to write well. Although we should not expect to achieve such success in writing well, we should remain committed to producing the best work possible.

CHAPTER 3

Scientific Publications

3.1   BASIC CONTENT

There are many different types of publications in science and engineering, including abstracts, conference proceedings, journal articles, books, theses, dissertations, and technical reports. We focus on that which is generally regarded as most important, however, the archival journal article. There are also different types of journal articles, including original articles, technical notes (sometimes called brief communications), and review articles. We focus on the original article, which is both most common and most important to the advancement of science and engineering, for it documents significant, novel findings. Some journals impose stringent guidelines on the organization of such an article, including particular subheadings, yet considerable flexibility often allows the author(s) to present the material in the best way possible. For purposes of illustration, however, we follow an outline recommended by the majority of scientific journals, namely

    Abstract
    Introduction
    Methods (or Materials and Methods)
    Results
    Discussion
    Acknowledgments
    References

Indeed, because most papers have the same basic structure, it is expedient to use a custom, generic file (e.g., called PaperTemplate.doc) to begin writing each paper. This file not only can remind us of the basic outline, it can ensure the proper formatting (often 1-inch margins, double-spaced, and 12-point font unless the particular journal states otherwise or provides its own template), including the placement of tables and then figures at the end of the manuscript. Having such an electronic outline in place can be a brief time saver, but perhaps, more importantly, it serves as a mental aid to begin writing.

Recall from Chapter 2 that there are five basic steps of writing well: formulating a detailed outline, writing freely, editing critically, reading out loud, and having a colleague critique the final draft. Moreover, a detailed outline not only includes major headings, as noted above for the original journal article, it also includes potential subheadings and either bullets that highlight key ideas or in some cases leading sentences. Although different authors employ different approaches in crafting the original detailed outline, a good place to start is to call all authors together and to lay out on a table the primary findings: figures, images, tables, equations, and so forth. These key findings can then be discussed and ordered logically, which will define the key bullets in the results and serve to remind us what methods were essential in obtaining the results. Once we have outlined the methods and results sections, it is then easy to outline the introduction and discussion. We discuss each of these key sections in detail below. First, however, let us consider a few items.

Submission of a paper for consideration for publication in a technical journal usually requires a cover letter to the editor, a list of potential reviewers, and a cover page. Let us begin with the cover page, an important part of the submitted paper.

3.1.1   Cover Page and Letter to Editor

The cover page serves to communicate to the editor and publisher a number of important pieces of information: the title (and thus subject) of the work, those who performed the work (i.e., the author list in a specific order) and their professional affiliations, keywords that classify further the area of study, and finally the full address of the corresponding author. The title is extremely important; it will determine to a large extent who reads the paper. A good title captures the essence of a paper without being overly long. Indeed, general rules of thumb are that the title should not exceed 120 characters and it typically should not contain verbs. Consider, for example, well-known titles from two of the most important and widely cited papers from the 20th century:

> On the Electrodynamics of Moving Bodies
> Molecular Structure for Nucleic Acids: A Structure for Deoxyribose Nucleic Acid

The first example is from Einstein's famous paper of 1905 that introduced his special theory of relativity; the second example is from Watson and Crick's famous paper of 1953 that introduced their concept of the double-helix structure of DNA. The *American Medical Association Manual of Style* (1998) recommends further that titles should not contain phrases such as "The Role of . . ." or "The Effects of . . ." or "The Treatment of . . ." and so forth. Although there is no need to be dogmatic when crafting a title, simple guidelines are useful reminders nonetheless.

Select keywords that are distinct from words used in the title and based on general, but not generic, aspects of the paper to ensure broader distribution. It is both much easier and more important today to identify appropriate keywords. One can log onto a standard computer-based search engine, such as PubMed, and compare results for different keywords to identify those that highlight papers most closely related to your work; such words would be good candidates for keywords. Because most investigators now search for technical papers using computer-based search engines, we cannot overemphasize the importance of appropriate keywords. In other words, writing well is not enough — if the work does not reach the intended audience, it will not have an impact. Significant attention must be given to the title, keywords, and, as noted below, the abstract.

Most journals require a cover letter even if the submission is completed online. This letter serves as the official "intent to submit" the paper and thereby to agree to all policies and procedures of review adopted by the selected journal. Among the many points that can be addressed in this letter, it is customary to confirm that the work is original, that the paper is not simultaneously under consideration for publication elsewhere, and that all authors contributed to the work and agree to its submission. One may also wish to note why the work will be of interest to the readership of the journal or to identify potential reviewers who either should or should not be selected and why. Like the paper itself, however, the cover letter to the editor should be concise. Here, we provide a simple example; letters will vary depending on individual circumstances.

Date

Dr. J. Smith
Editor, *Journal Name*
Address

Dear Dr. Smith:

Enclosed please find the manuscript entitled, "Title," which is submitted for consideration for publication in *Journal Name*. This paper represents original research that has not been, nor will it be, considered for publication elsewhere until a decision is reached by you or your staff. All authors contributed to the work and its preparation and agree to its submission.

With very best wishes, I am

Sincerely,

Name
Title

In some cases, the letter to the editor should also contain statements that, if applicable, all research involving human subjects or animals conformed to accepted standards and was approved by the appropriate institutional committee. Similarly, if applicable, this letter can communicate that permissions have been obtained from the appropriate parties to republish previously published work and it can offer suggestions of possible reviewers. In all cases, however, it is important, indeed often required, to include a statement that the paper has been submitted to only one journal for evaluation. That is, just as with submissions of grant applications to most agencies, there is a standing agreement among scientists and engineers that it is improper to submit the same work for simultaneous evaluation by two or more journals because of the significant effort required of others to provide proper and timely reviews.

Let us now turn our attention to the technical sections of a journal article. Rather than discussing them in the order in which they appear in a paper, however, we discuss them in the usual order of composition. Hence, we end with a discussion of the abstract rather than beginning with such a discussion.

3.1.2   Results

The section on results is the heart of the technical paper; it reports the primary findings, which often represent the most important information contained in the paper. The results should be easy to write, thus many authors prefer to write this section first. Indeed, in cases of multiple authors working on a single document, the first author often drafts the results and methods first, then the senior author drafts the discussion and introduction. All authors then revise the completed first draft. Regardless of approach, one of the best ways to write the section on results is to lay out all of the figures, tables, equations, or other major findings that you may include, then to prioritize and order them in the most logical fashion. It is important to emphasize in this regard that we need not order the results chronologically; in some cases, authors order results by importance. Once done, it is then easy to write freely. This approach is particularly effective when a paper is coauthored by two or more investigators, for laying the results out on a table facilitates discussion of the relative merits of each finding. Note, too, that although some journals require subheadings within the results, they are often omitted, and the lead sentence of each paragraph serves to introduce the different key findings. Indeed, some recommend that the lead sentence of each paragraph in the results should state the most notable finding in that paragraph.

One of the most frequently asked, and often most difficult to answer, questions is: How much interpretation of the findings should be in the results versus discussion? The reason that this is difficult to address is that it depends in part on the style of the author and recommendations by the specific journal. In general, however, most technical writers agree that the results section should be objective; it should focus solely on presenting the findings. Hence, although it is common to point out within results any interesting or important features within specific figures, images, equations, or tables, it is best to reserve for the discussion any interpretation of the significance of the finding as well as any comparison to findings by others.

Another question that arises often is how best to refer to a figure or table. For example, should we write

A and B were found to be related linearly (Figure 1),

or is it better to write

Figure 1 reveals a linear relationship between A and B.

In other words, is it best to state the key finding and refer parenthetically to the associated figure, image, equation, or table, or is it best to cite directly the particular evidence that reveals (shows, illustrates, or so forth) the key finding? Notwithstanding some exceptions (e.g., specific instructions to authors for some journals), the answer to this question is often that it is a matter of personal style. Note from our illustrative example, however, that the first approach involves passive voice, whereas the second approach involves active voice but further requires the figure, image, equation, or table to "do something" — reveal, show, illustrate, confirm, or so forth. Some editors suggest that inanimate devices such as figures cannot "reveal" or "show" such things, thus they prefer parenthetical references over the direct approach. Conversely, others prefer the crisp, active voice in the second example, which helps to minimize the use of passive voice as desired in general. We encourage the reader to consider these and similar options carefully and to adopt a consistent, but not rigid, personal style. Such a decision should affect other aspects of writing a technical paper, for example, the introduction wherein one often reads "This paper presents." Again, some would argue that only the investigators can present, not the paper, yet many prefer this crisp, active style of introduction.

3.1.3   Methods (or Materials and Methods)

The methods section is usually the easiest to write. Indeed, if one has trouble getting started in the "write freely" phase, it is often best to go to the methods. Simply put, the methods section is where we document how we accomplished the work. In principle, this section should contain enough detail to allow the reader to repeat the study in the same way — this alone allows one to test and confirm the basic tenet of science, reproducibility. Given the increasingly sophisticated methods and procedures used in science and engineering, however, writing an effective section on methods demands significant planning. Moreover, given that one can use many commercially available kits or software packages, there is a need for balance between detail and proper citation.

Two effective devices in writing the methods section are to use ample subheadings and parallelism. Here we emphasize that we should not write scientific papers to be read, we should write them to be studied. Subheadings thus aid the student in organizing information or locating quickly particular aspects of the methods when needed. Typical subheadings in a paper on cell biology could be:

Immunohistochemistry

In Situ Hybridization
Statistics

Subheadings in a paper on mathematical modeling could include:

Theoretical Framework
Constitutive Models
Numerical Methods

In either case, subheadings should proceed logically and thereby reveal to the reader the thought process followed by the investigators. The format for the subheadings, for example, numbering or italicizing, is dictated by the particular journal and thus is provided in the specific instructions to authors.

Because many scientific findings result from or imply a mathematical statement, it is important to address the treatment of equations within a paper, often within methods or results. Simply put, *write an equation as a normal part of a sentence*. For example, Newton's famous second law of motion is usually written simply as $f = ma$, where f denotes force, m denotes mass, and a denotes acceleration. Consistent with the presentation here, mathematical symbols usually appear in a distinct font, which may include italics (e.g., scalars) or boldface (e.g., vectors). In many cases, however, either for emphasis or simply because of complexity, equations are set off as a separate line within the text. In this case, the equation is still part of the sentence and thus should include commas and periods as appropriate. For example, we could write the following. It is important to remember that Newton's second law of motion, namely

$$f = ma,$$

holds only with respect to an inertial reference frame. Similarly, we could write the following. The governing equation in this case is Newton's second law of motion, which can be written as

$$f = ma.$$

An easy way to remember that equations are part of the normal grammatical structure is to recall step 4 from Chapter 2 on how to write well — reading out loud forces us to include equations as natural parts of the text. A final reminder is that most journals do not allow a nomenclature for symbols used within equations. Hence, one should always state the meaning of a symbol just before or just after the equation in which it is introduced, just as we did above for Newton's second law (e.g., noting that f denotes force).

For those symbols that are universally accepted or familiar to readers of a particular journal, there is no need to define them explicitly. Examples of well-recognized symbols are $+$, $-$, $=$, and also those for summation, derivatives, integrals, and so forth. Given the increasing complexity of science and engineering, commonly used symbols may represent multiple quantities within the same paper, hence there is a need for care. For example, R is often used for radius, but it is also used for the universal gas constant [$R = 8.314$ J/(g mol) K]. The most important suggestion in this regard is to be clear and self-consistent.

Finally, a frequently asked question relates to the level of detail needed in cases where one reports results obtained using methods reported previously in other journal articles. Although there is no rigid answer, the best practice is to document the essential, new methods and to refer the reader to previously established methods, where appropriate. For example, if your group established the previous methods, simply state that the details can be found in a previous publication, then briefly outline the methods; if others established the previous methods, cite the key paper(s) and provide a brief, but slightly more detailed summary of the methods. Conversely, one must provide significant detail when reporting a new method or procedure. Such detail can include specific instruments and vendors, chemicals and their sources and concentrations, specific versions of software, and so forth. In cases of human or animal research, one must first note that the appropriate institutional oversight committee (e.g., the Institutional Review Board, or IRB, for human research and the Institutional Animal Care and Use Committee, or IACUC, for animal research) approved the work.

3.1.4   Discussion and Conclusion

Most journals recommend against using a separate section for conclusions, which would typically be brief, hence the discussion often serves a dual role. One should address at least three specific points in the discussion:

> Interpret the specific results and emphasize the significance.
> Compare the current with past results.
> Identify limitations and potential needs for further research.

Whereas the introduction normally addresses the significance of the overall research topic or area, the discussion should address the potential significance of the particular findings. For example, an introduction may note the importance of cardiovascular consequences of hypertension, which affects more than 50 million Americans, but the discussion may note the significance of the new finding that blocking a particular receptor in vascular smooth muscle cells reduces hypertension in an experimental cohort. Like the introduction, therefore, the discussion should cite appropriate references, albeit often with greater discussion of the relevant details. It is important in this regard to cite only the most relevant literature. In other words, the goal is to place the current findings within the most appropriate context, not to provide an exhaustive collection of all previous work that is remotely related to the overall topic or specific findings. Because of the explosion of scientific and engineering knowledge, citing good review papers can often serve to cover general information without concern that some important papers may be missed. Related to issues of ethics, of course, one should not purposely fail to cite a relevant paper for personal gain.

A frequent question with regard to the discussion is how much information should be included on the inherent limitations or future needs. In some ways, this addresses both the style and the ethics of written communication. It is both prudent and useful to others to point out many of the limitations of the study, with justifications, for this will both put the current study in perspective and guide future work. Nevertheless, one must be careful not to focus on the negatives in a way that it distracts from the significant accomplishments or advancements of the study. The key, therefore, is to maintain an appropriate and candid balance. Similarly, it is useful to point the reader toward important, useful directions for future research. Yet, many investigators do this in a guarded fashion to allow themselves the opportunity to exploit their present findings and achieve further advances. The key point, therefore, is to maintain a proper balance — provide guidance so that others can advance the field while protecting intellectual property.

In summary, the primary goals of the discussion section of a paper are to reemphasize the significance or innovation of the study, interpret and discuss implications of the specific findings, compare the current findings with similar work by others, discuss limitations of the methods or findings, and summarize the major finding(s) while giving direction for future work.

3.1.5   Introduction

As with any introduction, the primary goal of this section is to capture the reader's interest and "set the stage." Toward this end, it is generally recommended that the introduction answer three basic questions:

> Why is the general topic or particular study important?
> What is currently known and what remains unknown?
> What does the current paper address or accomplish?

One should be able to answer these questions easily after having written the results. Consistent with answering these questions, the typical introduction consists of three to four paragraphs even though there is considerable variety in the number and especially the lengths of these paragraphs. Experienced writers may write the introduction first, but most writers write the introduction after completing the methods and results and sometimes even the discussion. Regardless, it is important to provide sufficient references in the introduction to justify both the need for the study and the general approach adopted.

An important issue with regard to writing a journal article, including the introduction, is the appropriate use of abbreviations. Good rules of thumb are to use only commonly known abbreviations (e.g., DNA), to use them only if the word or phrase is repeated three or more times throughout the document, and to introduce them at the first occurrence in the body of the paper (cf. Section 2.6.2). Some journals require the author to collect abbreviations together, for example, in a footnote on the first page of the paper or in a table. Regardless, it is best to use abbreviations sparingly.

3.1.6 Abstract

The technical abstract has always served an important role — it provides a brief summary of a paper and thereby helps a reader decide whether to read or study the paper. With the advent of computer-based search engines, however, the abstract has become a particularly important means of capturing the attention of the intended audience. Hence, albeit a short section, often not more than 250 words, the abstract deserves great attention.

Most writers compose the abstract last. It must reflect briefly the overall paper, including the basic motivation, significance, general approach, and key discoveries or final solution; it must be written clearly, without jargon, acronyms, or uncommon abbreviations, and must stand alone, generally without references. As with the introduction, the first sentence of the abstract should be engaging. In contrast with the introduction, the last sentence of the abstract generally summarizes the most important finding or points to pressing needs for future research. Whereas most journals allow the authors to write the abstract as they see fit, a few journals require the authors to follow a uniform outline, including specific subheadings. As in all cases of technical writing, it is thus important to read the "instructions to authors" for the particular journal.

3.1.7 Acknowledgments

It is customary, indeed required in most circumstances, to acknowledge the financial support that enabled the work. As an example, one might read:

> This research was supported, in part, by grants from the National Institutes of Health (R01 HL-10000 and R21 HL-01000).

Because of the increasing move in science and engineering toward translational research, investigators cite industrial support more frequently. In such cases, the journal may require the authors to disclose conflicts of interest related to potential financial gain related to the results. If there are no disclosures, this should be noted.

In addition to financial support, it is often appropriate to acknowledge technical support as well as intellectual or editorial contributions by individuals who were not involved extensively enough to merit coauthorship but who contributed nonetheless. Such acknowledgment should be merited, however, and those noted must be informed. Indeed, some journals now require individuals acknowledged in this section to stipulate in writing that they are both aware of and deserving of such recognition.

3.1.8 Appendices

Appendices are not found in all scientific papers; indeed, they appear in the minority of papers. Nevertheless, when used well, appendices can serve a very important role. In general, appendices contain important information that either does not fit well within the flow of the body of the paper or is simply best stated separately for those few readers who will be interested in such details. A good example of appendix material would be the step-by-step derivation of key equations, the final result of which can be found in the body of the paper. In this way, the author fulfills his/her responsibility of providing methods that are sufficiently detailed to enable the reader to reproduce a result while not distracting the reader from the key points presented in methods. Similarly, detailed "recipes" for molecular or cellular assays may fit well in an appendix.

3.1.9 References

It is interesting that we are often taught "proper methods of citation" in courses and books despite different journals and publishers requiring very different citation formats. In some cases, references must be arranged according to the order of appearance within the work and numbered sequentially beginning at 1; in other cases, references must be arranged alphabetically by the last name of the first author, then numbered beginning at 1; in yet other cases, references must be arranged alphabetically and not numbered. This basic scheme dictates how to cite any reference within the text — by number or by author. Similarly, the format for the references that details the authors, year of publication, title, volume, and inclusive pages also varies from journal to journal. The best advice, therefore, is to follow the specific instructions to authors. As specific examples, however, consider multiple ways to cite the same article within the text:

> Watson and Crick (1953) proposed the double helix . . .
> Watson and Crick [20] proposed the double helix . . .
> Watson and Crick[20] proposed the double helix . . .

> or similarly,

> The double-helix structure of DNA was . . .(Watson and Crick, 1953).
> The double-helix structure of DNA was . . .[20].
> The double-helix structure of DNA was . . . .[20]

As seen, the third approach in both cases results in some savings with regard to printing, which is important to some publishers given that *most journal articles cite ~35 papers* and most review articles cite over 100 papers. When these simple savings are multiplied 30-fold or more, one can appreciate the potential savings in page costs. It is noted, however, that numerical citation has the disadvantage that the reader must constantly refer to the reference list to determine who was responsible for the cited finding. Informed readers often know who has done what in a field, which is to say who has produced reliable or important findings. Citation by author names (e.g., Smith et al., 1999) thus has the advantage of increasing the flow of the paper. Nevertheless, one must follow the format prescribed by the journals and publishers. Usually the only case wherein one can pick a format is while writing proposals, which is discussed in Chapter 4.

Citation is similar to that discussed above when there is but a single author. For example, we might find the following: Einstein (1905) proposed . . ., Einstein [15] proposed . . ., or Einstein[15] proposed . . ., and similarly we might find . . .the special theory of relativity (Einstein, 1905), . . .the special theory of relativity [15], or . . .the special theory of relativity[15]. In the case of three or more authors, however, the format differs slightly. Recall from Chapter 2 that "and colleagues" is abbreviated in the Latin as "et al." Hence, we might find the following: Smith et al. (2008) proposed . . .or . . .(Smith et al., 2008). Whereas some journals use (Smith et al. 2008), that is, they omit the comma after the et al., it is a mistake to add a comma after the first author's last name (i.e., Smith, et al., 2008 is not an accepted format). Again, the best advice is to refer the specific instructions to authors for the journal of interest.

Finally, note that the citation within the reference list can also appear in various forms. For example, we can cite the same paper as:

Watson JD, Crick FHC (1953) Molecular structure of nucleic acids: A structure for deoxyribose nucleic acid. Nature 171: 737–738.

Watson JD, Crick FHC. Molecular structure of nucleic acids: A structure for deoxyribose nucleic acid. *Nature*. 1953;171:737–738.

Watson, J.D., and Crick, F.H.C., 1953, "Molecular Structure of Nucleic Acids: A Structure for Deoxyribose Nucleic Acid," Nature, **171** pp. 737–738.

Other formats exist, which is why one must consult the instructions to authors for each journal.

A final, and important, reminder is to cite only those papers that you have actually read. Perhaps surprisingly, many investigators will cite papers that someone else has cited simply because it is easier. Such a practice can be dangerous. In science and engineering, one should always check and double-check everything, including interpretations of other work used in citations.

3.1.10 Figures and Tables

It has been said that a picture is worth a thousand words. Actually, a well-prepared and appropriately selected picture (e.g., figure or image) can be worth a thousand words if done well. As an example, consider the following figure, a standard x–y scatter plot, which is the most common type of figure found in a technical paper. Although this example contains the basic ingredients of an effective figure (e.g., clear data points and labeled axes, with the unit of measurement denoted parenthetically on the x-axis), we can improve it considerably with little effort.

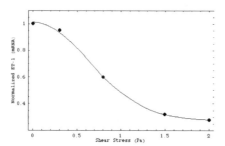

Compare the following version of this figure (reprinted with permission, CISM) to the previous one. It is easy to see that a reduced number of tick marks along each axis as well as larger numbers and lettering improve the readability considerably. Indeed, one of the most important considerations is that typesetters will reduce the size of many submitted figures before placement within the final version of the paper. This is particularly important when placing a figure within a single column in a dual-column layout, which most technical journals use.

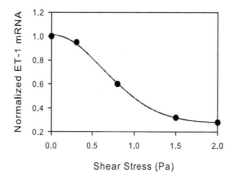

Albeit a small point, note that the solid curve in these figures represents a best-fit to data obtained using a formal regression method. Whereas solid lines are appropriate for showing such "theoretical" or "model" fits, it is best to use lightly dashed lines when the goal is to connect the data points for emphasis. It should be remembered, however, that simply connecting the dots may be misleading if data sampling missed key points.

Finally, realizing that many readers go to the results after having only read the title and perhaps the abstract, it is important to write complete legends so that the reader can understand the importance of the figure easily.

In summary, the original journal article should be both well written and well illustrated; it should address the following primary questions:

Introduction: What was done and why?

Methods: How was the work accomplished?

Results: What was found?

Discussion: Why are the results important, how do they compare to previous work, and what remains to be investigated?

Seeing one's name in print for the first time on an archival paper can bring a sense of excitement and pride. Seeing one's name on a paper that contains errors or fundamental flaws can bring a sense of regret. There is, therefore, a need to give such work our most careful attention from start to finish.

Exercise 3.1   Interview someone who serves on an editorial board of a technical journal and ask how reviewers are selected, what are good reasons for excluding reviewers in particular cases, and what is done when different reviewers have diametrically opposing views. Write a one-page summary of the results of the interview.

Exercise 3.2   All journals limit the number of words or pages allowed for papers within particular categories. For example, it is common for an original article to be limited to 6000 words inclusive. Noting that an abstract typically is ~250 words, each one-column table and a single panel figure is equivalent to ~250 words, and a standard full-citation reference is typically equivalent to 20 to 30 words (some up to 40 words), estimate the number of words for an abstract, 6 figures, 1 table, and 35 references, which is typical for a standard original paper. Next, calculate the number of words available and estimate reasonable lengths for remaining sections: introduction, methods, results, and discussion.

3.2   PUBLISHING AN ARCHIVAL JOURNAL PAPER

3.2.1   Origin

According to Boorstin, (1983, pp. 390–394),

> The printed scientific 'paper' or 'article', which was simply a later version of the letter, would be the typical format in which modern science was accumulated and communicated. . . .The letter was an ideal vehicle for the increasing numbers of men dispersed over Europe who no longer expected to storm the citadel of truth, but hoped to advance

knowledge piece by piece. . . . A letter had obvious advantages over a book. While works of science were often large tomes easy to stop for censorship, the novel observations in a letter could slip unnoticed or be delivered with the 'ordinary post'.

In contrast to early European investigators such as Galileo (1564–1642), few modern investigators need be concerned about potential censorship of their work. Rather, the primary concern today is to ensure that a paper receives broad distribution to the intended audience. Toward this end, electronic publishing and the Internet have revolutionized the availability of scientific papers, yet the methods of composition, presentation, and submission have not changed.

3.2.2   Composition and Authorship

It is difficult, if not impossible, to write by committee. Indeed, one of the most important documents in American history, the Declaration of Independence, was assigned to a committee of five for composition but was drafted in seclusion by a single author, Thomas Jefferson. As it should have done, the committee evaluated and revised the final draft penned by Jefferson before forwarding the final version for consideration by the Continental Congress of 1776.

Although there has been a significant increase in the number of authors on scientific papers, particularly in biomedical science and engineering, the primary responsibility of writing a paper must similarly fall to one author or in some cases two authors (e.g., the first and senior authors). As noted above, however, the best way to ensure that the first draft represents the ideas and expectations of all authors is to meet together to define the initial outline and to discuss what findings to report in the results. We address the issue of joint authorship further in Chapter 6, hence we merely note here that it is essential that all authors agree on the contents and presentation of a paper before submission for consideration of publication.

3.2.3   Submission and Review

As noted earlier, the essential first step when preparing a paper for consideration for publication is to read the instructions to authors for the intended journal. Only in this way will one be able to fulfill the requirements of each journal. In general, however, the two primary items needed for submission are the aforementioned cover letter to the editor (cf. Section 3.1) and the complete paper, including the cover page, body of the paper, tables, and figures.

Exercise 3.3   A few journals allow authors to submit a paper for consideration directly to a member of the editorial board or the sponsoring society. In these cases, that member can assume the sole responsibility for review and may then "communicate" the paper to the editor for publication. Identify two different journals that allow such a procedure and write a two-page summary discussing the history of this approach and the associated advantages and disadvantages.

An editor or associate editor will usually solicit two or three experts to provide a recommendation on the potential suitability of a paper submitted for publication. These reviewers are asked to provide objective assessments and thus to decline to review a paper if there is either a real or a possibly perceived conflict of interest. The period allowed for review varies considerably among different journals, with some biological and clinical journals allowing only 2 weeks and some mathematical journals allowing up to 3 months for review. Differences also exist between journals with regard to the possible categories of recommendations available to the reviewer, but general categories are:

Accept (i.e., accept as submitted)
Accept pending minor revision (not requiring rereview)
Major revision (with required rereview for further consideration)
Reject (i.e., not suitable for publication)
Inappropriate for this journal.

In cases where the topic, type, or length of a paper is deemed to be inappropriate for a journal, an editor can communicate this to the author(s) without a formal review, even though reviewers are also allowed to make such a recommendation independently. A recommendation to reject a paper may be rendered for any of a number of reasons: the paper does not contain original or novel findings, it contains serious flaws in design or analysis of the data, it does not address a relevant problem, results contradict previously published findings without addressing adequately the associated reasons, there is insufficient new information (i.e., it is only an incremental advance at best), and so forth. In the case of a recommendation to reject, the editor should ask the reviewer(s) to state this case diplomatically, although this does not always happen. It is common for a journal to reject 50% or more of all submissions.

The recommendation to request major revisions generally implies that there is a need for additional experiments, analysis of data, or computations. Such a recommendation can also reflect the need to correct a major, but not fatal flaw, to reduce the length of the paper if it is overly long, or to improve significantly the presentation, including improved figures, tables, and writing.

In contrast, minor revisions (not requiring further review) can reflect a need to expand the methods or discussion or conversely to eliminate information that is available in other publications. There may also be a need to add some key references or reduce the overall number of references. Minor improvements in writing or the need to eliminate results that are duplicated in tables and figures may also lead to a recommendation of the need for minor revisions.

Finally, albeit very uncommon, the recommendation to accept (as is) suggests that the study is important, novel, and well presented. All authors should strive to submit such work, but should be prepared to revise a paper as needed.

As one might expect, unanimous recommendations by two to three reviewers are uncommon, hence the editor or associate editor usually must make a decision based on the information received.

For example, if two reviewers recommend major revisions and a third reviewer recommends rejection, the editor is justified in recommending either major revision or rejection. In cases where the editor rejects a paper, the authors can try to rebut the reviews and request either the privilege to submit a revised paper or that additional reviewers are asked to review the paper further. In most cases, however, editors tend to stand by the initial, carefully weighed decision, and it is best to consider other options. For example, some authors will simply resubmit the same paper to another journal for consideration for publication; in such cases, they are usually not required to reveal that the paper has been rejected previously, which enables the second assessment to be objective. Authors must realize, however, that reviewers are often picked carefully for their expertise, and it is possible that different editors from different journals will select the same reviewers. For this reason, and simply because one should always use any opportunity to improve a paper based on any feedback obtained, it is best to revise a paper that has been rejected before submitting it to another journal.

Finally, it is useful to know how editors select reviewers and what instructions are given to the reviewers. Ideally, the editor, associate editors, or editorial consultants are familiar enough with the topic of the submitted paper that they know the experts personally or at least know of them. In cases wherein such experts either decline to review a paper (because of conflicts or simply because they are not able to provide a timely review) or cannot be identified easily, editors will often peruse the references cited in the paper. In other words, frequently cited authors are good potential reviewers because their work is related closely to that which was submitted and has passed previous peer review. Alternatively, editors may also use computerized search engines to identify potential reviewers based on the publication of similar work in reputable journals.

3.2.4 Revision

As noted earlier, only a small percentage of technical papers are accepted upon the first submission, hence authors should expect to revise a submitted paper. Indeed, in most cases, revision along the lines suggested by the reviewers improves the paper significantly, thus *revision should be seen as an opportunity not a failure*. Nevertheless, it is human nature to be disappointed or, in some cases, upset by a negative review. Toward this end, we recommend two things. First, read the review carefully but do not take any action until at least a few days later. In other words, neither a rash response to the editor nor a hasty attempt to revise a paper is likely to be productive. Second, avoid both of the two most natural responses to a negative review — to adopt all of the reviewer's recommendations because "they must be the expert" or to ignore all of the comments because you "know better." Rather, it is best to take comments and concerns by reviewers at face value. For example, if a reviewer states that a particular section is hard to understand even though you think it is clear, chances are that at least some other readers will also find the section hard to understand. The best response in this case, therefore, is to take advantage of the opportunity to improve its clarity and to ask a colleague to assess the changes. In other words, because it is your name on the paper, take every opportunity to make the paper the best it can be.

When a journal allows or requests a revision, the author(s) usually must submit the revision within a certain period (often 3 to 6 months, but highly variable) and provide evidence of the revisions. In some cases, the author(s) can meet this requirement simply by summarizing the revisions on a separate page or in a letter to the editor. In other cases, the journal may either require separate detailed responses to each of the reviewer's concerns or marks within the submitted manuscript that identify the revisions. The latter requirement is now met easily using features such as "track changes" in MS Word.

Once the authors decide to revise a paper, they should ask how to do this most efficiently. For major revisions, it is best to identity the requisite experiments, calculations, analyses, and so forth and to generate the additional results. Next, one can follow the same approach used in writing an initial draft — lay out all results, new and old, on a table and determine which to include and in which order. Once done, revise the results, methods, discussion, introduction, and abstract accordingly and document the revisions appropriately. For minor revisions, it can be efficient to begin by writing the required "response to reviewers" or "summary of revisions." After knowing how one wants to address the concerns, the paper can then be revised accordingly.

Although policies differ among journals, it is uncommon to allow authors to submit more than one revision because of the time invested by both the editors and the reviewers. In other words, one should work very hard to satisfy reviewer's concerns and to make a revised manuscript acceptable.

3.2.5   Typesetting, Galley Proofs, and Proofreader Marks

Years ago, a typewritten copy of a manuscript was copyedited, then typeset from scratch. Today, nearly all journals require electronic versions of accepted papers, which are then copyedited and formatted for publication. Briefly, copyediting is an important step wherein a manuscript is checked carefully by the editorial staff of the journal, or a third-party associate, for format, style, spelling, complete and consistent citations, and so forth. In some but not all cases, the copyedited version is sent to the authors along with the galley proof to show the changes that were made.

A galley proof is the final draft of the paper formatted exactly as it will appear in print. The authors are asked to check the galley proof carefully to ensure accuracy, yet it is expected that only minor changes or corrections will be made at this stage of publication. In the case of major changes due to author errors, the publisher may charge extra to make the requested changes. For this reason, authors should ensure that the final version of a manuscript is correct as submitted. Although the advent of electronic publishing decreased tremendously the number of necessary corrections, authors should be diligent to check the galley proof carefully, including the layout of tables, figures, and equations. If errors are discovered after approval of the galley proof and publication of the paper, the authors' only recourse is to publish an erratum (or errata, which is the plural form of erratum and used in the sense of correcting simultaneously multiple errors).

Because of the importance of the galley-proof step, publishers developed a nearly universal set of symbols and directives to communicate changes that were needed in the paper. Again, however, because of the increasing capabilities of word processors, it is now common for such corrections to be made electronically — for example, using the "comment" function in PDFs or the track-changes function in word files. If needed, however, one can find proofreader marks in standard dictionaries (e.g., *American Heritage Dictionary*) or online.

Exercise 3.4   Find a 10- to 20-page paper that you have written previously and scan it for minor errors. Use standard proofreader marks to note the appropriate changes just as you would do for a galley proof.

3.2.6 Copyright, Permissions, and Page Charges

Copyright is a legal procedure that grants exclusive rights to the production, publication, sale, or distribution of a work by the owner of the copyright. Because this deals with the ownership of ideas, it is addressed in Chapter 8. Here, therefore, we simply note that upon acceptance of a paper for publication, the publisher will usually request that the authors transfer copyright to the publisher. Although copyright agreements tend to be standard for the publishing of scientific papers, one should read such agreements carefully before signing. If there are questions regarding anything within the agreement, one should either consult a more experienced author or contact the copyrights division of the publisher.

Transfer of copyright requires that the material to be transferred is original and, when appropriate, that special permissions have been obtained to republish any previously published material. In the latter case, the most common situation is the desire to republish a figure or image from another work. One may obtain permissions to do so by writing the copyrights division of the publisher that holds the copyright and requesting permission to republish the work in a specific way. In most cases, a publisher will grant such permission provided that a simple statement is associated with the republished material [e.g., "From Smith (2001), with permission from *Publisher Name*."]. In some cases, however, such permission is granted only after receipt of a fee, which could be hundreds of dollars for a single figure. In cases of financial constraints, it is best to contact the publisher early to identify potential fees. Although one only needs to contact the holder of the copyright, which is often the publisher, not the original author(s), requesting permission from the author, if he/she can be located easily, is a good gesture.

In some cases, one step remains before the publication of your paper — payment of "page charges" or fees to cover the publication of color figures. In many cases, journals charge standard fees, according to the length of a paper, to offset portions of the cost of publication. Such page charges are mandatory in some cases but voluntary in other cases. The rationale behind voluntary page charges is that many agencies that financially support research also desire that the associated findings be published, and consequently, they provide researchers with funds to support publication. Payment of page charges often entitles the author(s) to free hardcopy reprints or PDF versions of the paper. Independent of page charges, some journals assess mandatory fees for the publication of color figures in print (but not electronic) versions. Because fees for color figures can be hundreds to thousands of dollars, it is prudent to minimize the use of color figures or images to those that are best understood in color. With the continued growth of online publishing, however, one can also consider using color figures in the online version that are equally clear if printed in black and white by the reader or by the publisher of the print version. Regardless, it is also good to determine before submission if such fees will be charged. Like all submitted figures and images, color versions must have the required resolution and must be submitted in the proper file format. Again, the author is referred to the instructions for authors for each journal for requirements vary with journal.

Exercise 3.5   Read the article "Structural Outline of an Archival Paper for the *Journal of Biomechanics*" [Brand RA, Huiskes R (2001) *J Biomech* 34: 1371–1374]. Construct a two-column table to record hints therein that reinforce or contradict that which was presented in this section (3.2). Provide a summary, not to exceed one page, that articulates your preferences in cases where there is disagreement.

3.3   THESIS OR DISSERTATION

Most universities require the completion of a thesis as part of the requirements for a master of science (M.S.) degree or a dissertation for the completion of a doctor of philosophy (Ph.D.) or a doctor of science (D.Sc.) degree. It is common for an M.S. thesis to range from 50 to 150 pages and for a Ph.D. or D.Sc. dissertation to range from 100 to 250 pages (each double spaced with ample margins). Although it may seem a formidable task to write such a long document, it is actually not difficult if one formulates a good outline and simply writes chapter by chapter. One should obtain specific guidelines for formatting such documents from the local Office of Graduate Studies, however, for requirements differ from institution to institution. Here we note briefly the two most common styles for organizing a thesis or dissertation.

First, a thesis or dissertation can be organized along traditional lines and thus consist of the following:

Abstract
Chapter 1: Introduction
Chapter 2: Background
Chapter 3: Methods
Chapter 4: Results
Chapter 5: Discussion
Chapter 6: Conclusions and Recommendations
References
Appendices

Note the strong similarity between this outline and that for the archival journal paper, with two notable exceptions. To encourage students to understand the literature well, most institutions require a separate chapter, entitled background, which highlights past work on the topic of graduate study and identifies areas of need for further study. Because a dissertation must represent original work, review of the background literature is a particularly important part of the doctoral student's early work. To encourage students to recognize the limitations of their own work and thereby to identify areas of further study, most theses and dissertations end with a chapter entitled "conclusions and recommendations." A simple way of thinking about recommendations is to ask the question, "What should the next graduate student in the laboratory do to push the current work forward?"

In contrast to the traditional outline, it is becoming increasingly common to organize theses and particularly dissertations around individual papers that are based on the student's work and either have been or will be submitted for publication in archival journals. For example, consider the following outline:

Abstract
Chapter 1: Introduction
Chapter 2: Paper 1
Chapter 3: Paper 2
Chapter 4: Paper 3
Chapter 5: Conclusions
Appendices

Packaging the multiple papers (often one for a thesis and three to seven for a dissertation) between general introductory and concluding chapters allows the student to focus on writing the individual journal papers. In this case, each chapter contains its own introduction, methods, results, discussion, and references. It is clear that the individual papers should be written first, as described above, and the introduction and conclusions written last. In the traditional case, it is common to write the background first, then methods and results, and finally discussion, introduction, and conclusions and recommendations. If one adopts this second style of organization and papers are published before submitting the final dissertation, it is important to consider issues of copyright; the best advice in this case is to check with your local office of graduate studies.

3.4   TECHNICAL REPORTS

Whether a technical report will be published or not, it is a formal document and thus requires careful attention. In contrast with the aforementioned types of technical documents, however, the technical report does not demand a particular outline. Indeed, such reports can range from a one-page summary to a thousand-page document. The best advice, therefore, is to discuss in detail the expectations before beginning the outline, then use, as appropriate, the aforementioned guidelines for writing an archival journal paper.

Exercise 3.6   Find and read carefully the instructions for a thesis or dissertation at your institution. Write a one-page bulleted summary of the key stylistic requirements.

CHAPTER 4

Proposals and Grant Applications

4.1 INTRODUCTION

Whether it be a proposal to undertake a particular senior design project, a graduate student's proposal to a committee to pursue a particular area of research for his/her thesis or dissertation, an employee's request of management to support a new area of R&D, an organization's application for state or federal support, or a professor's request for support of basic research, the technical proposal is fundamental to securing the resources needed to advance science and engineering. Notwithstanding the many different types of proposals, most are similar in basic structure and preparation. Hence, for illustrative purposes, we focus on the NIH individual investigator grant application known as the R01, the primary mechanism for funding health-related research in the United States.

4.2 TYPES OF GRANTS

The R01 mechanism supports two general types of grant applications: investigator-initiated and those in response to a request for proposals (RFP) or program announcement (PAR). By investigator-initiated, we mean a scientist or engineer identifies a fundamental question or important problem, conceives of an approach to address this issue, and takes the initiative to request the funding needed to complete the project. In contrast, an RFP or PAR is a "call for proposals" that addresses a particular need or area of investigation that a working group, committee, or administrator has deemed important. In the latter case, the scientist or engineer must first learn of the opportunity, then respond according to the stated instructions. Indeed, one of the most important aspects of a successful grant submitted in response to a call for proposals is that it is responsive to the announcement.

Exercise 4.1   Go to the NIH Web site, www.nih.gov, and search for current PARs. Pick a PAR of interest and read the instructions carefully. Write a three-page summary of the PAR that would be sufficient as an overview of the motivation, scope, and requirements for submission.

Exercise 4.2   The National Science Foundation (NSF) funds basic research and training in the sciences and engineering. Go to the NSF Web site, www.nsf.gov, and search for current PARs. Pick an announcement of interest and read the instructions carefully. Write a three-page summary that would be sufficient as an overview of the motivation, scope, and requirements for submission.

There are three general classes of motivations for any proposal: hypothesis-driven, curiosity-driven, and technology-driven. No one motivation is more important or more scholarly than another; they are simply different.

Hypothesis-driven: The NIH defines a hypothesis as "an educated prediction about the outcome of your study." Under some programs, the omission of a hypothesis is a major oversight, one that can result in the reviewers suggesting that the proposal is merely a "fishing expedition," that is, a project without clear direction.

Curiosity-driven: We have all heard the saying that when asked why he/she climbed the mountain, the climber simply stated "Because it was there." Curiosity-driven research is the desire to answer a new and intriguing question. Curiosity has been, and will remain, the primary driver of scientific inquiry.

Technology-driven: In our increasingly technology-based society, there are many cases wherein the ability to design and build a new instrument is motivation enough to pursue such research. Indeed, in many cases there are excellent opportunities to modify previous designs to address new applications.

Regardless of the motivation — hypothesis, curiosity, or technical need — the approach to apply for and secure funding for research is similar.

4.3   THE REVIEW PROCESS

Proposals undergo a two-step review process at the NIH. First, proposals are evaluated for technical merit and feasibility. Second, they are evaluated administratively for funding potential. A committee called a study section, under the direction of a scientific review administrator (SRA), accomplishes the first step; a council accomplishes the second step while attempting to balance the desire to fund the best science and the need to accomplish the fundamental mission of the NIH — to improve the health of people in the United States.

Just as we should identify the intended audience before writing a journal article or giving an oral presentation, we should also understand the audience that will read and review a particular proposal. In contrast to many funding agencies, the NIH publishes the names of those constituting most study sections, which enables the applicant to know the audience. In this case, it is prudent to read recent publications by members of the study section to get a feel for their scientific interests and basic perspectives. Although each member of the committee will be asked to score all grant applications for which they do not have a conflict of interest, only three to five individuals generally read each application in detail. These individuals are referred to as the primary reviewer, secondary reviewer, and discussant(s); they are selected based on the closeness of their technical expertise to that of the proposal, as revealed primarily by its title and project summary. Recalling that we use the NIH R01 herein mainly to illustrate issues that are important in preparing a proposal, a graduate student would similarly be well advised to know the composition of his/her committee and to read recent papers by these individuals to anticipate what types of questions might arise.

Whether or not one knows the composition of a review panel, perhaps the most important things to know are the criteria that the panel will use to evaluate the proposal. Again, we use the NIH criteria, but the need to be familiar with such criteria would even apply to a master's or doctoral research proposal, which a faculty committee would evaluate during a proposal defense. At the NIH, the evaluation criteria include[1]:

Significance: Does the study address an important health problem?
Approach: Are the design and methods appropriate to address the aims?
Innovation: Does the project employ novel concepts or methods?
Investigators: Are the investigators appropriately trained?
Environment: Will the scientific environment contribute to success?
Overall evaluation: Will the study advance health care or medical science?

Additional considerations of importance during a review include whether a proposed human or animal research protocol respects current guidelines; before beginning any such study, a local institutional committee (e.g., IRB or IACUC) must evaluate and approve such protocols, but this need not be completed before submission. Similarly, reviewers will address whether the proposed budget is justified. We focus here, however, on the scientific aspects of the review enumerated above. Indeed, because reviewers are asked to comment specifically on how well an applicant addresses these criteria, it can be helpful to the reviewer, and thus the applicant, to include succinct, highlighted statements that address significance and innovation in particular.

Because each reviewer can interpret what is meant by significant, innovative, or appropriate, it should not be surprising that different reviewers often have very different opinions with regard to the value of the same proposed research. Hence, the applicant should try to write in a way that engages different people, noting in particular that not all reviewers are experts in the specific area proposed. Rather, it is expected that good scientists and engineers can recognize good work when they see it; a key challenge, therefore, is to help the reviewer appreciate the overall objective, significance, logic of the methodology, and innovation of the proposed work from both a general and a problem-specific perspective. This may be even more important in cases where the reviewer has

---

[1]Portions of this discussion are based on *Proposal Writing: The Business of Science* by Wendy Sanders, then at NIH (www.wm.edu/grants/PROP/sanders.pdf).

expertise in the proposed area but a different scientific opinion. In this case, the need is even greater for convincing, objective arguments.

There are, however, a number of things that all reviewers appreciate. For example, because reviewers are often asked to review multiple proposals (e.g., 8 to 12), it is important to *make the proposal easy to read and understand*. Rather than using a small font or small figures to fit more information into the same space, use larger fonts and ample figures to aid the reader and write clearly and concisely to increase the value of what is provided. Indeed, this brings us back to the importance of Chapter 2. Note, therefore, that in a 1987 publication of the NIH, entitled *Helpful Hints on Preparing a Research Grant Application to the National Institutes of Health*, one reads

> Try to develop a clear, concise, coherent scientific writing style. A few guidelines that may prove useful include: (1) using the active voice, which is more direct, less wordy, and less confusing than the passive voice; (2) keeping together related ideas and information, for example putting clauses and phrases in sentences as close as possible to the words that they modify; (3) simplifying and shortening overly long and involved sentences and paragraphs; and (4) eliminating redundant and awkward words, phrases, and sentences.

Additional, subtle things can likewise make it easier for the reviewer. For example, although citing the many references numerically (often 100 or more) can save space and thus enable the applicant to include additional information, this forces an educated reviewer to refer back continually to the references to determine what work has been cited. It can be constructive, therefore, to cite references by name, for example, Smith et al. (1999) or (Smith et al., 1999) rather than [20], for the reviewer may know of and respect the work of Smith and colleagues. Similarly, recall that schematic drawings, images, figures, and tables can each be worth a thousand words if done well. Indeed, many reviewers will first skim through an application by paying particular attention to these visual aids. For this reason, it is useful to provide descriptive legends so that the reviewer does not have to search the text to find the meaning and importance. Because of the importance of the accompanying text, however, it is also good practice to cite the schematic drawings, images, figures, and tables using boldface type to enable the reviewer to locate easily that part of the text that discusses each figure (e.g., boldface **Figure 1** and **Table 1** are much easier to identify quickly in the text than are Figure 1 and Table 1). Although they should be kept to a minimum, introducing key abbreviations in boldface also enables a reviewer to find them much more quickly in the text, for example, nitric oxide (**NO**) versus nitric oxide (NO). The key, therefore, is to keep the reviewer's perspective in mind at all times and not to compromise the use of effective devices and strategy because of stated limitations on the number of pages allowed – concise writing will generally provide the extra space needed to include all the necessary information.

Exercise 4.3   The NIH Web site, www.nih.gov, has specific instructions for writing a K-series grant application. Write a three-page summary that would be sufficient as an overview of the motivation, scope, and requirements for submission of such a grant application to the NIH.

4.4   THE NIH R01 GRANT

As noted previously, the R01 grant is but one of many funding mechanisms administered by the NIH. One can obtain information about the other types of grants from the NIH Web site (www.nih.gov), but we consider here the format for the R01 because it represents well how to design an effective application.

The R01, or single investigator grant, consists of a cover page, brief description (project summary) and list of primary personnel, table of contents, budget and budget justification, a biosketch for each of the primary personnel, information on resources (i.e., the research infrastructure) that are available to the investigator(s), the main body of the application, and further administrative information. With the exception of the main body of the application, all other information must be provided within appropriate NIH-supplied form pages. Again, see the NIH Web site for instructions and details on the overall grant application package, including form pages.

Here, we focus on the main body of the application, that is, the five basic sections that detail the scientific need and proposed method of approach. These sections are: specific aims, background and significance, preliminary results, research plan, and references. Whereas the R01 application currently allows 25 pages, single-spaced, for the main body of the application, other types of NIH applications have different requirements. One of the most important aspects of successful grant writing is to follow the instructions, which includes respecting page limitations and using approved fonts and margins. For example, the current NIH R21 mechanism allows 15 pages, single-spaced, for the main body of the application, whereas the NIH BRP (Bioengineering Research Partnership) mechanism allows 40 pages, single-spaced. Whether 15, 25, or 40 pages, a key to successful grant writing is to write with clarity and conciseness. Indeed, after having written numerous 25-page applications, we have found that trying to provide the same level of detail in a 15-page application is a very good exercise — it forces one to write more concisely.

Before discussing in detail each of the five primary sections of the main body of the grant, note that each section should answer a specific question:

Specific Aims: What are you going to do?
Background and Significance: Why is it important?
Preliminary Results: Are you capable of being successful?
Research Plan: How are you going to accomplish the work?
References: What are the key findings on which you will build?

As in any good technical document, the writing should flow logically from section to section and the applicant should reinforce the main ideas throughout. Similar to the situation of multiple authors writing a paper, when multiple investigators write different parts of a proposal, it is important for the principal investigator to ensure a consistent style throughout, including tense.

4.4.1   Specific Aims (1 Page Required)

The specific aims section is particularly important; it must capture the reader's interest, show the need for the proposed research, and detail specific results that will be sought — all in one page. Different applicants use different formats, but a general approach to constructing the specific aims is to begin with two or three short paragraphs that identify the overall problem or long-term goal of the applicant(s) as well as the specific problem that needs to be addressed and why, then list the individual specific aims, and conclude with a brief paragraph that highlights the innovation and overall significance of the work.

There is no limitation on the number of specific aims that one can propose, but most R01 applications focus on three to four aims, which sometimes include multiple sub-aims. Just as it can be efficient to begin writing a technical paper by identifying the primary findings, so too it can be efficient to begin writing a proposal by first identifying the specific aims. Indeed, it is often useful to draft the specific aims page and have multiple colleagues provide feedback on the overall plan before beginning to write the remainder of the proposal

Specific aims should be just that, specific. Moreover, construct these aims in a forceful way — to quantify, to determine, to design, to prove, to develop, and so forth. Many applicants construct each specific aim to test a specific hypothesis; the key objective, therefore, is that the aim is testable. Although one can provide some indication in this first section as to how the aims will be accomplished (e.g., using a particular animal model or data from a particular clinical trial), it is best to focus on methods and approaches in the section on Research Plan.

4.4.2   Background and Significance (3 Pages Recommended)

In some ways, the background and significance section can be the hardest section to write well. Whereas one may think of this section simply as a brief literature review and statement of the obvious (e.g., that the problem is significant because a particular number of Americans experience the highlighted disease or problem), it actually must be much more. Within the context of answering the question, "Why is this research important?," the applicant should critically assess the literature to show convincingly what is unknown and why this lack of understanding is impeding scientific advances, improvements in health care delivery, the development of better medical devices, and so forth. In the words of the NIH, there is "a need to identify the gaps" in our knowledge. For example, we may know that a genetic mutation is responsible for a particular disease, but we may not understand how this mutation affects the activity of a particular type of cell. Similarly, we may know that hemodynamic factors give rise to a particular vascular pathology, but we may not know how the associated forces induce the changes in gene expression that ultimately cause the disease. Although identifying gaps will often require one to point out shortcomings in previous investigations of others, we should do this diplomatically.

When identifying key gaps in the literature within the background section, one should show convincingly the need for the proposed specific aims. In other words, background should "set the stage" for the research plan. In addition, however, we must remember that not all reviewers will be intimately familiar with the specific area of research, hence also use this section to educate the reviewer so the he or she can appreciate better the importance of the identified gaps and the proposed method for research. For example, if the study seeks to identify the relative effects of particular cytokines in a disease process, some background on cytokines — their discovery, general activity, half-lives, receptor affinity, and so forth — may help the reviewer appreciate the motivation for the underlying hypothesis. Well-illustrated schematic drawings, flowcharts, figures, and images often add considerably to this section.

Finally, remember that this section is entitled background and significance. If one commits three pages to this section, only half of one page will typically be devoted to significance. Nevertheless, significance is one of the criteria that the reviewers must address, and the council increasingly bases funding decisions on significance. It is useful, therefore, to address in this section the potential impact of the overall project, that is, the importance of filling the identified gaps. To aid the reviewers, the applicant should highlight key points in this section, for example, by italicizing, underlining, or boldfacing the text. Given the overall importance of significance, but limited space in this section, successful applicants are generally very good at weaving the significance throughout the proposal: the last paragraph of specific aims, the significance portion of background and significance, and the rationale sections in research plan. The challenge, therefore, is to reinforce key points throughout without redundancy.

4.4.3 Preliminary Results (6 Pages Recommended, But Not More Than 9)

The primary goal of the preliminary results section is to demonstrate the capability of the principal investigator(s) and assembled team to accomplish that which is proposed in research plan. In many ways, this is the easiest section to write; indeed, if you are having difficulty getting started, it is often good to focus first on the preliminary results.

This section is best written in the style of a results section in an archival journal paper, but with abundant subheadings. Proceed logically by documenting your previous successes on closely related problems (with citations to previous journal publications) or your new results that show explicitly that you can successfully complete the proposed aims. It is beneficial, therefore, to remind the reviewers when such results demonstrate that a specific aim can be achieved or that hypotheses on which they are built are tenable. Like a good results section in a journal paper, this section should contain copious figures, images, equations, and tables that highlight key findings. One of the key challenges can be the decision on how much information to include from past publications because the reviewers have access to such information if so directed. The best advice is to provide sufficient detail on critical methods or findings — that is, make it easy on the reviewer by not forcing him or her to find and read the previous paper — but to refer the reader to original papers for nonessential information that nevertheless may strengthen the argument. Indeed, whereas the NIH used to allow the applicant to deposit up to 10 key previous papers in an appendix, this is no longer possible if the paper can be found on the Web. Just because a paper can be found on the Web does not mean that a busy reviewer will take the extra time to do so, however; again, it is critical to make the most important information readily accessible within the grant application itself.

Because of the importance of team science in biomedical research, most grant applications to the NIH rely on a team of collaborators. Hence, it is also important to demonstrate the capability of the different investigators to work well together just as they will need to do during the proposed project. The best way to demonstrate this is via joint publications, or at least joint abstracts for papers presented at technical meetings. In the absence of such evidence, it is important to show that materials, data, and so forth have already been shared as will be required by the proposed research.

4.4.4   Research Plan (15 Pages Recommended, But Not Fewer Than 12)

Recall that the primary question that needs to be answered in this section is, "How are you going to accomplish the proposed work?" In conjunction with the specific aims, this section is the most important and thus demands careful attention. Perhaps the best word to remember when writing this section is "detail."

There is no required format for research plan, yet an effective strategy has evolved over the years. Many applicants begin this section with a paragraph that highlights the overall research plan and its importance, sometimes including a schematic drawing to show how the different aims complement one another. Next, they describe the rationale, methods, and expected results/potential limitations for each aim in sequence. Finally, they conclude with a brief summary of the overall project and an expected timeline to accomplish the project. One small variation on this strategy has also arisen in recent years, due in large part to the extensive but common procedures used in molecular and cell biology. Similar to the format of some technical journals, one can collect detailed methods (often common to multiple aims) at the end of this section, almost like an appendix, so as not to interrupt the flow of the main portion of the section; this allows the interested reader to evaluate the appropriateness of the details nonetheless. Indeed, in some cases, these detailed methods are set apart by the use of a smaller font, which saves some space while emphasizing the importance of the preceding text.

If one adopts the most common strategy, then the basic outline for the main portion of this section becomes[2]:

Aim 1. Restate the specific aim exactly from the first page.
Rationale.
Methods.
Expected Results and Limitations.
Aim 2. Restate the specific aim exactly from the first page.
Rationale.
Methods.
Expected Results and Limitations.
Aim 3. Restate the specific aim exactly from the first page.
Rationale.
Methods.
Expected Results and Limitations.

Restating each aim in its entirety reminds the reviewer of the specific goal — to quantify, to determine, to design, to prove, and so forth. Restating the aim in boldface serves as a natural and effective visual cue for organizing this long section; this approach is much less distracting than needless section numbers such as 4.1, 4.1.1, and so forth. Whereas the significance section discussed previously should focus primarily on the importance of accomplishing the overall project, the rationale should focus on the fundamental reason(s) for each specific aim. For example, the applicant should note what important gap in our understanding this specific aim will address and why the adopted approach is innovative.

The methods section for each aim is similar to a methods section in an archival paper — it should provide methodological details sufficient to enable the reviewer to repeat the study. For

---

[2]There are many slight variations, however. For example, one could replace the generic methods section with separate sections on experimental design and data analysis, or one could separate the expected results and limitations section and rename the latter potential difficulties and alternate approaches.

example, one should not write "the cells will be cultured in an appropriate media." Rather, one should document the specific media to be used (including vendor), any supplements with appropriate concentrations, and the temperature and CO_2 level. Similarly, one should not write "the governing differential equation will be solved numerically" or "the data will be analyzed for possible statistical significance." Rather, one should provide details on the specific numerical method and why it is appropriate for the expected class of differential equation, and similarly one should provide details on the specific statistical tests, including post hoc testing, why they are appropriate, and the levels of desired significance. Again, the operative word in this section is *detail*, assuming of course that the methods are both appropriate and proven.

Although NIH-funded research does not need to be hypothesis-driven, one should always anticipate the results. It is thus prudent to discuss why you expect such results, which actually allows one to justify further the importance of the aim. Although results should be new, it is always good to cite related studies that provide further confidence that the aim will prove successful and important. Similarly, although methods should be chosen and justified carefully, it is always possible in science and engineering for difficulties to arise that prevent one from conducting the experiments or analyses as originally planned. There is also a need, therefore, to anticipate such difficulties and to have reasonable contingency plans. Just as in the discussion of an archival journal paper, however, one must achieve the proper balance in identifying potential pitfalls while not implying that the aim will be very difficult to achieve as planned. It is wise to discuss the presentation of this balance with a valued colleague.

Finally, it is useful to conclude the research plan with a detailed timeline showing the anticipated duration of each part of the project and how the different parts will progress together, perhaps in different laboratories. It is also good to provide a single paragraph that concludes the grant — *remind the reviewer what the key gaps are in the literature and how the present study will fill these gaps using innovative approaches that promise significant findings.*

4.4.5   References

The reference section was limited to four pages in the past, but there is currently no such page limitation. Nevertheless, one should not seek to compile an exhaustive list of references; it is more important to be selective, focusing on the key papers that support the need for the research and the methods used to address this need. Similarly, there is no required format for references except that each must include the list of authors, year of publication, title of the work, the publisher, or journal title, volume, and inclusive pages. Because reviewers are typically familiar with the proposed research area, and thus the key authors in the field, it can be helpful to list the references alphabetically. Indeed, this is consistent with the aforementioned recommendation to cite by author (e.g.,

Smith et al., 1999) rather than by number (e.g., [20]), for this eliminates the frustration felt by knowledgeable reviewers who do not want to go back and forth to the references to see who did what. Because of the availability of research papers through the Web, some applicants also provide links to enable the reviewers to download key references easily.

4.5 THE PREPROPOSAL

Perhaps the best example of the need to write concisely with clarity is the preproposal. Because of the greater numbers of applicants applying for limited financial resources, many agencies have instituted a two-stage review process. The applicant must first submit a brief preproposal, which an expert panel will review. Based on the findings by this panel, only a subset of full proposals is invited for consideration for funding.

The state of Texas, for example, has a competition called the Advanced Research Program (ARP) that is open to any full-time member of the faculty of a Texas institution of higher learning. Preproposals for ARP grants have been limited to 4000 characters (use the word/character counting feature of your word processor to count); this is essentially $1^{1}/_{3}$ pages, double-spaced, in 12-point font — not a lot of information. Yet, a panel will decide whether to invite the applicant to submit a full proposal, the next important step toward possible funding, based solely on these 4000 characters. Again, the need for clarity and conciseness is clear. The format for the ARP preproposal is simple:

- Project goals and methods
- Staff
- Facilities and resources
- Education and training

Although such preproposals are very short, one clearly wants to communicate information similar to that contained in the much longer R01 application: What are you going to do? Why is it important? Are you capable of being successful? How are you going to accomplish the work? Recalling these simple questions, and noting four sections required for the proposal, it is prudent to think carefully how to partition the essential information within the required format. For example, whereas one uses the preliminary results in a NIH application to demonstrate capability, it may be better to use the section entitled staff in the ARP application. Similarly, whereas it may be appropriate to highlight the available equipment in preliminary results or research plan in an NIH application, it may be more appropriate to list these in facilities and resources in the ARP application. Again, the key thing to consider when beginning a grant application is what information you feel will best represent you and your ideas. Only then will you be able to decide best how to package this information within the format for the particular agency.

4.6   SUMMARY

The Whitaker Foundation recently closed, but it provided millions of dollars of funding over decades to support new investigators in biomedical engineering and to develop new academic programs. They provided reviewers of individual investigator grants with a checklist to ensure that applicants covered a number of critical aspects of research in biomedical engineering. Reasons for scoring a Whitaker application poorly included:

No clear hypothesis
Mundane/uninteresting
Little engineering
Little biology
Not enough detail
Unrealistic/faulty approach
Needed collaboration missing

Other reasons commonly cited for scoring NIH applications poorly include:

Not significant; not innovative; not exciting
Unjustified hypotheses
Unaware of previous related work
Insufficient pilot data
Poorly designed research plan; unorganized
Overly ambitious
One or more aims are poor
The success of one or more aims depends on the success of a previous aim

Although we should focus on the positives, it is prudent to appreciate causes for failure. In summary, some of the most important reminders for grant writing are:

Know the mission of the agency and target the proposal accordingly. For example, you would not think of sending a proposal on cancer research to the American Heart Association.
Read a recently funded proposal to the agency to which you are applying.
Read the instructions and follow them carefully when preparing your application.

Ensure that the proposal addresses an important issue and offers the potential for significant advancement.

Remember that your proposal must generally address simultaneously two technical audiences: those who are very familiar with the field and those who are less so.

Finally, finish early so that colleagues can review the application and provide constructive criticisms that you have time to employ. Only in this way can we avoid the common pitfalls that plague so many proposals.

Exercise 4.4   Write a 4000-character preproposal using the Texas ARP format. Select a topic of interest to you, assume you are the only investigator, and describe resources available in your laboratory or department that would be sufficient to conduct the work.

Exercise 4.5   The NIH Web site (www.nih.gov) provides useful guidelines on "How to Write a NIH Grant." Go to the site, review the material, and prepare a 25-slide PowerPoint presentation that could be used as an introduction to writing and submitting NIH grants.

APPENDIX (Copy and use this as a quick reference)

Specific Aims (What you are going to do?)

- The first sentence or two should engage the reader and motivate the need for the work.
- Briefly note long-term goals/overall hypotheses, then draw focus to the work.
- State your specific aims (three to four) and how you will achieve/test them.
- Conclude by emphasizing the novelty and innovation of the proposed work.

Background and Significance (Why it is important?)

- Review the literature critically, that is, identify foundations and gaps. Do not simply state that A did this, B did that, and C did that. Gaps are important areas that your work will address and fill.
- Being unaware of important findings in the field does not engender confidence; conversely, citing work from recent meetings or personal communication with leaders in the field suggests that you are on the cutting edge (do not overdo though).
- Only a few of the reviewers will have expertise in the specific area, yet many will read the proposal. Background should educate the general reader.
- Significance refers to overall importance and long-term potential rather than the significance of each of the aims — address the latter as rationale in research plan.

Preliminary Results (Are you are capable of being successful?)

- This section should do two things: demonstrate capability in the area and demonstrate feasibility with respect to the specific aims. In other words, convince the reader that you are capable of successfully completing the aims as stated.
- A picture can be worth a thousand words, and so too a table, flowchart, figure, or equation — illustrate the proposal well, taking note that an aesthetically pleasing document that is easy on the eyes is much appreciated. Use many subheadings while avoiding the use of numbering sections (e.g., C.1.1, C.1.2) that simply forces the reader to think about a nonessential.

Research Plan (How you are going to accomplish the work?)

- One of the most effective strategies is to address each aim separately, but to do so in a consistent, well-ordered manner. For example, for each aim, cover in subsections (a) rationale, (b) methods, (c) expected results and limitations.
- The rationale of each aim should address the importance of this part of the project and how it fits into the overall/long-term goal. This is also a good time to remind the reader of novelty or innovation. One short paragraph should suffice.
- Methods for each aim may include materials, equipment, theoretical frameworks, assays, statistical methods, and so forth, all given in sufficient detail. For example, do not merely say that a physiological solution will be used — give the specific composition. Similarly, do not just say that a particular device will be used — give the resolution of the device and any unique capabilities.
- Whether hypothesis- or curiosity-driven, one should know what to expect with regard to findings. Discuss this and note its potential importance. Likewise, one should know what limitations or pitfalls may arise. Noting and addressing them is much better than hoping a reviewer will not think of them; someone always does and this could relegate an otherwise outstanding proposal to a lower score.
- Finally, remember that *detail* is the operative word in research plan and that the aims should form a logical, supporting sequence. Tell them what you are going to do, how you are going to do it, and briefly why it is important.

CHAPTER 5

Oral Communication

Just as we must write well, so too we must speak well — a belief that is not new to modern science or engineering. According to Boorstin, (1983, p. 395), Bishop Sprat suggested that the goal of the Royal Society of London (founded ~1660) was "not the Artifice of Words, but a bare knowledge of things." Hence, they

> extracted from all their members, a close, naked, natural way of speaking; positive expressions; clear senses; a native easiness: bringing all things as near the Mathematical plainness, as they can: and preferring the language of Artizans, Countryman, and Merchants, before that, of Wits, or Scholars.

In other words, as Boorstin concluded, "It was not enough that the language of science be simple. It had to be precise — and, if possible, international." Although audiovisual aids available today are very different from those of the 17th century, the need for simple, clear, and informative presentations remains.

Written documents and oral presentations both reflect one's professional reputation. Yet, the oral presentation is unique in that it can serve as the all-important "first impression." If a talk is lucid and enjoyable, those in the audience will likely seek out the speaker again; if a talk is poorly organized and boring, it may be the last time that they seek to hear the speaker.

Exercise 5.1 The need for excellence in oral presentations is not unique to science and engineering. Hence, find a good book on public speaking and read two chapters that are particularly appealing. Write and submit a three-page summary of the main points. Among the many books available, consider the timeless work, *How to Develop Self-Confidence and Influence People by Public Speaking*, by Dale Carnegie.

5.1 EFFECTIVE STYLES

Carnegie (1956) suggests that four things are essential in one's pursuit of becoming an effective public speaker:

1. Start with a strong and persistent drive.
2. Know thoroughly what you are going to talk about.
3. Act confident.
4. Practice, practice, practice.

Although these four essentials should not surprise anyone, they should cause some reflection. In particular, just as in writing well and ensuring integrity in the workplace, effective oral presentations do not just occur, even with experience — *one must resolve to learn to present well and to continue to improve*. Moreover, because of the importance of self-confidence when speaking to either small or large audiences, it is essential to know the subject so well that you could give the talk even if the audiovisual equipment failed or if you forgot your typewritten notes. Finally, the old adage "practice makes perfect" is certainly true, but there is one caveat. One can practice a bad talk over and over, but it need not improve. Rather, one's practice should include peers constructively criticizing both the technical material and the method of presentation; it is better to make mistakes among friends and to receive helpful suggestions or corrections before the actual presentation.

Exercise 5.2   Attend three professional seminars and record seven specific personal habits and seven audiovisual techniques used by the speakers that were particularly effective (four each) or ineffective (three each). Summarize your findings in a table and submit a two-page report.

Many suggest that much of communication is nonverbal during discussions between individuals. Do we look the other person in the eye and reveal our interest or do we look at other people or things while they are talking? Do we change our facial expressions appropriately to reveal sympathy or understanding or do we remain stoic? So too in public speaking, nonverbal communication can help make a talk engaging or it can render the attempt boring or, even worse, annoying. By definition, habits are natural and repetitive; they usually arise unconsciously and can manifest nonverbally or verbally. For this reason, it is essential to have peers provide feedback on potentially distracting habits that arise while we speak. For example, if one tends to jingle keys in his pocket when nervous, recognizing this problem allows him to remove the keys before speaking, thus removing the potential distraction. Similarly, if one uses a lot of *aaahs* or *uumhs*, there is a need to identify these problems and remove them from both formal and informal speech, for we develop new habits through consistency. Indeed, note that we have found that paying careful attention to composing well-written documents also serves to help us speak well. Finally, if one's hands shake badly during a talk, it is best not to use a laser pointer, which will project exaggerated motions onto the screen. Instead, one whose hands always shake should practice using verbal cues such as "as seen in the first term of Equation 1" or "as illustrated well in the top curve in the left panel." A laser pointer can be an effective aid if used well, but it can also be very distracting. Indeed, even if held by a steady hand, a rapidly moving or constantly circling laser pointer can be a significant distraction. Finally, be careful not to keep the laser on if you "talk with your hands," for the audience gets both distracted and concerned when the laser shines across someone's face or constantly goes from floor to ceiling.

Valiela (2001) correctly suggests that effective technical presentations share some commonalities with successful theater. Two prerequisites for good theater are a good story and actors who "connect with" or "relate well to" the audience. A good story in science or engineering requires an interesting or important problem to be formulated, then solved in a novel and logical manner. Below, however, we focus on relating the story well to an audience, first by tabulating reminders related to basic techniques and habits of effective presentations. Indeed, although it is essential in science and engineering to have something important to say, as you compare the suggestions below, consider the suggestion of Carnegie (1956) that "It is not so much what you say as how you say it."

DO	DO NOT
Be confident, appear confident	Be arrogant or prideful
Be enthusiastic — it is contagious	Pace too much
Speak loudly, clearly, slowly	Speak in a monotone voice
Be respectful of questions	Ask rhetorical questions
Finish early enough for questions	Go over the allotted time
Maintain balanced eye contact	Look only at screen or at a distance
Dress appropriately	Apologize for dress
Use (laser) pointer effectively	Circle everything with laser pointer
Know your audience	Discourage interactions
Define terms, use analogies	Use jargon, try to impress
Minimize nervous habits	Assume every talk begins with a joke

Exercise 5.3   Why is it important to be or at least appear confident? Why is it important not to be prideful or arrogant? What message will we convey to an audience if we finish early and allow questions? What message will we convey if we go over the allotted time and ignore calls to stop?

What is the appropriate dress for different audiences? Ask yourself these and similar questions regarding this tabulated list of things to "do" and "not do," and write a two-page summary. If possible, conclude the summary with a few overarching statements.

Experienced actors tend to be nervous on opening night, and so too experienced speakers tend to be nervous before walking up to the podium. Yet, recognizing that nervousness is natural, indeed expected, allows us to identify ways to minimize its effects and to settle into a comfortable rhythm quickly. For example, an early visit to the room where you will speak will help you to feel more at ease — the environment will not be foreign. If you need to use a microphone and have not done so before, ask the A/V technician if you can test the system before your talk. If you expect to be nervous nonetheless, eat sparingly before the talk to avoid further complications of the nervousness. Having complete command of the technical material will also engender self-confidence, which is the best way to negate nervousness. Remembering the first sentence or two will ensure a good start, which is essential in transitioning from nervousness to confidence. Beginning with an engaging slide will capture the audience's attention, which will reinforce your confidence (provided you make eye contact with the now engaged audience). Conversely, memorizing a talk word for word can promote nervousness; you may become concerned that you will forget something and lose track of your message. Below we tabulate some reminders related to basic techniques that promote confidence as well as contribute to telling the story well.

DO	DO NOT
Visit the room before speaking	Show up late or just before your talk
Remember the first sentence	Memorize the talk or read it directly
Use an engaging first slide	Start with a bulleted outline
Use slides as your reminders	Require audience to read a lot on own
Maximize good figures/images	Use lots of words and small fonts
Be consistent in slide format	Mix slides with different backgrounds
Use slides to capture attention	Use slides to communicate most info
Remember the concluding remarks	End by saying, "Well that is all I have"

Reflecting on these suggestions, it should be clear that we recommend that one *use audio-visual aids to support a talk, not to carry it*. In other words, the speaker should strive to capture the audience's attention so that they look at him/her and only look to the slides when so directed for clarification. Hence, the speaker should use comments like "and thus x is important, as illustrated well in this figure" or "x . . . as can be seen in this image," noting that a well-used pointer can remind the audience when and where to look. Conversely, detailed text on a slide will usually entice the audience to read on their own and not to look at or listen to the speaker; this situation should be avoided. Use slides primarily to show clear black and white or color images and figures, schematic drawings and flowcharts, equations, and to a lesser degree, tables, each of which should support what is said. Providing a short heading on each slide can indicate the focus of that slide; beginners may also put bullets on the slides as further reminders to themselves, particularly to prompt appropriate transitions. When referring to figures, start by defining the variables of interest and the axes; when referring to equations, start by defining the meaning of important variables or terms; when using color images, use the different colors as indicators of important features or points. Remember, too, that *less information explained well is always better than more information explained poorly*.

Software programs such as PowerPoint can be tremendous tools when used well. Resist the temptation, however, to use all the "bells and whistles." For example, having figures fly in from the edges of a slide or animating molecules that come to screeching stops generally distract from the technical content. Similarly, using complex backgrounds, particularly ones with gradients in color, can be less effective overall — some words show up well, while others do not. Remember, too, that some members of the audience may be color blind; appropriate choice of color, particularly when delineating curves in figures, must be given careful consideration. Depending on the fixed lighting in the room, slides having dark backgrounds can excessively darken a room and thereby create a more conducive sleeping environment. For these and other reasons, black print on a white background and color images on a white background continue to be effective for they generally project well, maintain modest lighting in the room, do not discriminate unnecessarily against color blindness, and even allow one to use information directly from print versions of abstracts, proceedings, or papers that often appear in primarily in black and white because of considerations of cost.

The first slide is traditionally a title slide — it should give a brief (60 to 120 characters) but informative title and list the authors and their affiliations. Many try to add a touch of color by showing the university or business logo or perhaps a picture of a building or scenic area in which the group works. The last slide is traditionally an acknowledgment slide — it should list others who contributed to the work, financial support, and relevant disclosures. Some prefer to read the names and the funding agencies, but it is sufficient simply to list them in most cases. The last slide often remains projected the longest, that is, during the question and answer period, thus it is also a good place to list key references to your work and to provide contact information (e.g., an e-mail address). Consider adopting a common format/master slide, which enables you to use these slides for different talks with minor modifications; having a common format (including font sizes for headings versus text) enables you to insert any slide from a different talk into the present talk with no modification.

Finally, the next to the last slide usually provides a summary of the work or the "take home" message. It is best to end on a high note, emphasizing the major findings, rather than listing all of the limitations or future needs. Address such needs in response to appropriate questions.

Exercise 5.4   Prepare a 15-minute PowerPoint presentation on effective grant writing. Practice the talk, paying particular attention to the time limit. Have two or three peers critique the presentation, then make corrections and repeat the presentation.

Exercise 5.5   Prepare a 15-minute PowerPoint presentation on a technical topic of your choice, but do so in a way that that highlights bad presentation skills and personal habits. For example, use different backgrounds from slide to slide, use small fonts, use long detailed quotes, read directly from the slides, and so forth. Exaggeration often provides an important reminder of what not to do.

Exercise 5.6   Prepare a 15-minute PowerPoint presentation on a topic of your choice that addresses an issue having potential ethical consequences. For example, previous students in our classes have discussed embryonic stem cell research, cloning, the use of human subjects in clinical trials, animal research, issues of science and religion, patents, and copyright.

5.2   THE 15-MINUTE PRESENTATION

Seeing your name appear in print on a journal article generally produces a sense of accomplishment and pride. So too, learning that your abstract or paper has been accepted for a podium presentation at a national meeting produces a sense of excitement. After the initial euphoria, however, you realize that you have to find a way to describe in a short period, often 12 to 20 minutes, a project that you may have worked on for months or years. One is tempted, therefore, to pack as much information into the talk as possible. Surely the audience will be impressed by how much you did, right? As noted above, however, you will generally make a much more positive impression if you present less information well. There is, therefore, a critical need to identify the most important information and to ensure a logical sequence from identifying the problem to interpreting the results and appreciating the significance. Similar to writing a technical paper, a good way to start this process is to collect together all of the figures, images, equations, tables, or other major findings that you may include, then to prioritize and order them in the most logical fashion. This ordering need not be chronological; in many cases it is best to order the talk in the way that makes the most sense in hindsight.

A good rule of thumb is to prepare approximately one slide per allowed minute of presentation, including the first (title) and last (acknowledgment) slides, which need not be discussed. Moreover, each slide should generally highlight one main idea. Again, we emphasize that the first slide after the title slide should capture the audience's attention. It is much more effective, for example, to show a picture or image that motivates the work than to show a bulleted outline noting that you will introduce the overall problem, describe some of the methods, discuss the results, then draw conclusions — one expects such an approach. Carnegie (1956) suggests multiple ways to capture the audience's attention immediately: "arousing curiosity, relating a human interest story, beginning with a specific illustration, using an exhibit, asking a question, opening with a striking quotation, showing how the topic affects the vital interest of the audience, or starting with a shocking fact."

A brief anecdote highlights the importance of the second slide (or first slide when one does not use a title slide) in a PowerPoint presentation. One of the authors was asked to give the second technical talk at an anniversary celebration for the college of engineering. The first technical talk followed directly some brief comments by the president of the university. Out of courtesy, the president remained for the first talk because it began immediately following his comments. During the subsequent question and answer period, however, the president discretely moved toward the rear of the auditorium. Yet, as he approached the door, it was evident from the podium that the first slide of the second talk had captured his attention — the talk began with "This electron micrograph shows the fine structure of the heart and in particular. . . ." The president remained standing at the door and listened to the entire 10-minute talk. It is very important to capture the audience's attention quickly.

Finishing well is equally important to effective presentations. The conclusion is often that which the audience remembers best. Although Carnegie (1956) wrote on public speaking in general, not technical communication, it is interesting nonetheless to consider his suggestions for ending a talk: "summarizing, restating, outlining briefly the main points you have covered; appealing for action; paying the audience a sincere compliment; raising a laugh; quoting a fitting verse of poetry; using a biblical quotation; building up to a climax." Regardless of approach, ensure consistency between the opening and closing and try to memorize the ending so that it is thoughtful and forceful. Remember, too, that two of the best words to end with are "thank you."

It is important to embrace the question and answer period. Although many speakers tend to abhor criticism and do not want to be questioned, one can obtain valuable suggestions and guidance during this exchange. Indeed, many times, one will learn something that will improve the quality of a subsequent paper that will be written on the topic of the presentation. Three useful guidelines are: first, repeat the question both to ensure that you address what was really asked and to help the audience hear both question and answer; second, be respectful even if the questioner is antagonistic or if the question is truly a "dumb" question; and third, if you do not know the answer to the question, say that you do not know. It is best, however, not to answer all questions by stating that you do not know, hence the need for complete command of your subject. Finally, if a questioner tends to be unrelenting, suggest that you would enjoy discussing the issue at the next break. Remember, too, that because the question and answer period can be illuminating to both the speaker and the audience, finish the presentation early to allow sufficient time for this important exchange.

Although we addressed only the typical 15-minute talk, presentations of other durations should be treated similarly. Indeed, if you become proficient at "telling your story" concisely, it is easy to do so for any specified duration. The one caveat, however, is to remember that you should always present concisely — a longer duration simply means that you should communicate more information, not that you should communicate the same information less well. Remember, too, that it is always good to think ahead about which slides to skip if time is running out or if there is an unavoidable delay or slowdown. In other words, be prepared and be flexible.

Exercise 5.7   Prepare a 30-minute technical presentation, on a topic of your choice, using approximately 30 slides. After having given the presentation to your peers, reduce the presentation to 15 minutes without losing any significant technical content.

5.3   SUMMARY

It has been said that "Everyone has but one story to tell, they merely tell it in different ways to different people." We reemphasize, therefore, that one must know the intended audience. Rather than trying to sound scholarly, it is most important to be clear and effective. Avoid jargon; define terms carefully; read faces in the audience to obtain a sense of their understanding and engagement.

Recall from Chapter 1 that individual differences can bring a freshness and vitality to a field; individual personalities can generate excitement and interest. Each person should develop a style that is most effective and natural for him or her. The guidelines presented in this chapter are simply that, guidelines. We encourage the reader to consider or try the ideas presented here, but more importantly, to pay close attention to styles and techniques used by different speakers in different settings. You will be well served to take note of what is most effective and what is most ineffective, and to adjust your style accordingly.

CHAPTER 6

Authorship

Seeing your name on your first published paper may be one of the most exciting moments in your career. Many students thus enter a discussion on authorship focusing on the question, "How do I get my name on a paper?" In our experience, the more important and difficult questions include when and how to keep your name off a particular paper and how to negotiate questions of authorship among your collaborators in a multi-investigator project.

Exercise 6.1 If you have authored a journal article, answer the following questions about your experience before proceeding. If not, interview someone who has authored a journal article and report their answers.

1. How did you become an author on your first paper?
2. What was your contribution to that paper?
3. Who decided whose names would appear and in what order?
4. At what point during the research did you first discuss authorship?
5. Did you sign a legal agreement as an author, and if so, to what did you agree?

6.1 THE SLUTSKY CASE

Many widely publicized cases of research fraud, plagiarism, and other forms of misconduct exist in science and engineering. Discussing these cases often sheds light on important aspects of ethics in science and engineering. We will take as an example the case of Dr. Robert Slutsky, a member of the faculty at the University of California–San Diego School of Medicine in the 1980s. While in many ways similar to other cases of plagiarism or data fabrication, the Slutsky case is unusual for two reasons: the university committee, formed to investigate allegations of research fraud against Dr. Slutsky, included a philosopher as well as medical school faculty, and the committee attempted to draw broader conclusions about this type of fraud. The committee ultimately published its findings in an article in *The New England Journal of Medicine* (Engler et al., 1987). We briefly review details of the case below, but this excellent article is so integral to our discussion that it should be read before proceeding.

Exercise 6.2   Read the journal article regarding the Slutsky case [Engler RL, Covell JW, Friedman PJ, Kitcher PS, Peters RM (1987) Misrepresentation and responsibility in medical research. *N Engl J Med* 317: 1383–1389] and list the five aspects of the case you find most surprising:

1.

2.

3.

4.

5.

Dr. Slutsky, then associate professor of radiology at the University of California-San Diego, was being evaluated for tenure when a member of the tenure committee noticed an apparent duplication of data in two published research papers. The ensuing investigation by a faculty committee revealed a number of striking facts of interest for our discussion. First, the committee found clear evidence that Dr. Slutsky reported fictitious experiments and statistical analyses, reported incorrect procedures and statistical analyses, and listed colleagues as coauthors who did not contribute to the work and in some cases did not know about the publications. Second, the normal peer review process did not detect any of these concerns in the fraudulent papers, and some of the journals refused to retract the fraudulent papers upon notification of the committee's findings unless Dr. Slutsky agreed to the retractions. Third, at one point during the period under investigation, Dr. Slutsky was publishing a paper every 10 days, including many in prestigious journals. Fourth, much of this work was apparently sound; the committee established the validity of 77 of the 137 publications they reviewed and classified only 12 publications as fraudulent. Finally, the investigation revealed missed warning signs over the course of Dr. Slutsky's early career: several of his colleagues and at least one journal editor questioned the validity of manuscripts they read, and some recommendation letters for his original appointment to the faculty expressed concerns about the validity or quality of his research.

6.2   BASIC CONVENTIONS

Before discussing common problems regarding authorship, it is helpful to review current conventions. These conventions will be familiar to practicing scientists and engineers but not necessarily to undergraduate and graduate students, particularly those who have not yet authored a journal paper.

6.2.1   Order of Authors

The order of authors on an archival journal paper usually has special significance, but conventions vary by field and occasionally by journal. In most biomedical science and engineering journals, the first author is usually the one who performed most of the work; this person is often a graduate student or postdoctoral fellow who worked on the project described in the publication. Designation as first author is so important that footnotes are sometimes used to indicate equal contribution by two or more "first" authors. The last author is typically the senior investigator who conceived, guided, and financially supported the project. The ordering of all other authors is generally of less significance, as we assume that their contributions were less but otherwise important. In stark contrast, some fields encourage an alphabetical listing of authors. Notwithstanding customary variations by field and journal, surveys of scientists and engineers reveal widespread disagreement and confusion regarding conventions for authorship (Bhopal et al., 1997; Tarnow, 1999).

6.2.2 Submission Agreement

Most journals ask the author(s) to sign a submission agreement. Typically, this agreement transfers copyright to the publisher and asserts that the author(s) will pay any page charges levied by the journal as part of the publication process. In addition, this agreement usually asks the author(s) to verify the accuracy of the submitted manuscript and that it has not been published by or submitted to another journal. Much more variable are policies regarding who must sign the agreement. In many cases, the corresponding author (i.e., the person who submits the manuscript and lists his/her contact information in the final version) can sign on behalf of all coauthors. This explains, in part, how Dr. Slutsky submitted some papers without the knowledge of some people he listed as coauthors. Conversely, some journals require all coauthors to sign the submission agreement; it appears that Dr. Slutsky subverted this requirement by forging the signature of some coauthors (Engler et al., 1987).

6.2.3 Publication Impact

One's record of publication is critically important when applying for jobs, grants, awards, or promotion and tenure. In any discussion of authorship, it helps to understand how reviewers evaluate your published works. Obviously, one important factor is the number of publications, but this is far from the only consideration. For example, some journals are more selective and more widely read than others; publications in these journals are typically valued more in an evaluation. Such assessments are subjective, however, because investigators in the same field may have different opinions on the relative quality of the relevant journals or their ability to assess the quality of a particular work. For example, a complex mathematical model of a biological process will likely receive a more rigorous review by a mathematics journal than by a biology journal even though the latter may have a larger readership. In an attempt to weigh the quality of a journal more objectively, one can define quantitative metrics. One such metric is the "impact factor," a measure based on the idea that more frequent citations of a journal's articles implies a greater impact by that journal on its field. Scientific information service companies, such as Thomson Scientific, compute and report impact factors for a wide range of journals (to obtain links, search for "Journal Citation Reports" or "ISI Web of Knowledge"). These services also track the number of citations of particular publications, and some reviewers use the number of citations as a surrogate measure of the impact of a publication.

Even if one relies on a metric such as impact factor to value a publication, most publications have multiple authors. The question then becomes, "How much credit should each author receive for a given publication?" For example, consistent with conventions discussed above regarding the order of authors, the first and last authors typically receive most of the credit for any biomedical publication. When someone evaluates your publication record, they will notice not only the number of publications and the quality of the journals but also how often you appeared first or last in the author list. Note, however, that even if your name appears last on a publication, implying that you were the senior author most responsible for the ideas, if a well-recognized senior colleague also appears on the paper, other scientists may assume that your senior colleague deserves much of the credit for the ideas. This issue becomes especially relevant in multi-investigator collaborations, which are more common in today's research.

6.3   COMMON PROBLEMS

In June 2005, an article in the journal *Nature*, titled "Scientists Behaving Badly," reported results from a survey of more than 3000 NIH-funded scientists regarding the frequency with which they engaged in a range of questionable research practices (Martinson et al., 2005). While only 0.3% admitted to falsifying research data within the previous 3 years, many more admitted to some of the other problems highlighted by the Slutsky case, such as publishing the same data in two or more publications (4.7%). Particularly relevant to our discussion is that 10% admitted inappropriate assignment of authorship within the past 3 years. Strikingly, such misbehavior was more common among mid-career scientists (12.3%) than early-career scientists (7.4%).

Exercise 6.3   Read the article "Scientists Behaving Badly" [Martinson BC, Anderson MS, de Vries R (2005) *Nature* 435(9): 737–738] and formulate three hypotheses why mid-career scientists are more likely to engage in admittedly inappropriate behavior than early-career scientists. Compare and discuss your hypotheses with a colleague.

6.3.1   Expectations

The importance placed on publications as a measure of career progress can create substantial pressure to publish, particularly for tenure-track junior faculty. Managing this pressure begins by developing clear and reasonable expectations.

Exercise 6.4 For Ph.D. students and postdoctoral fellows. Answer the following questions regarding the number of publications you expect a junior faculty member to produce in your field. First, estimate the number of publications one might produce (or you did produce) during doctoral study. Next, estimate the number of publications one might produce (or you did produce) during 3 years of postdoctoral research. Finally, estimate the number of publications a successful junior faculty member in your field should produce during the first 5 years of his/her career. Now, perform two different "reality checks" on your estimates:

1. First translate your estimates of productivity into rates (number of papers per year, which may be < 1), noting that most papers tend to be produced near the end, not beginning, of one's study. Then, use these rates to compute how many graduate students and/or postdoctoral fellows your model junior faculty member would need to employ if each paper was coauthored by only one student or fellow. Do you think these numbers are reasonable? Which of your estimates would you adjust based on this check?
2. As a second check, ask a senior faculty member in your department to give the same three estimates. How do they compare to your estimates? If possible, discuss any discrepancies with a senior colleague.

One thing that seems apparent in the case of Dr. Slutsky is an unrealistic expectation (or perception of external expectations) regarding productivity. No reasonable person expects a junior faculty member in any field to produce a paper every 10 days. Yet, Dr. Slutsky apparently felt pressure to improve upon the number of valid publications (at least 77 in 7 years according to the authors of the report in the *New England Journal of Medicine*) through various types of research fraud. Clearly, there is a need for open discussion of authorship and productivity with everyone involved, from students to advisers to department chairs. Only in this way can we develop and clearly express realistic expectations regarding the number and quality of publications.

6.3.2 Gift, Guest, and Ghost Authorship

Gift authorship entails granting authorship to a person who did not contribute directly to the work (Davidoff, 2000). As an example, a new trainee discovered upon her arrival in Dr. Slutsky's laboratory that she was an author on a paper she knew nothing about (Engler et al., 1987). Why would someone do this? Misplaced generosity could be one motivation — colleagues may believe they are doing you a favor by listing you as an author on a publication. Another possible reason could be the pressure to show productivity by trainees who are supported by certain types of grants. Regardless, gift authorship could associate your name with a fraudulent paper, as in the Slutsky case. Cases of research fraud are rare, however; embarrassment is a more likely concern if you consider the paper to be of poor quality, you disagree with its conclusions, or you are forced to admit (e.g., during the question and answer period after a scientific talk or during a discussion with a colleague you respect) that you did not contribute to the work.

One of the most difficult situations related to authorship is receiving an unwanted gift authorship, especially if you are a junior colleague of the person conferring it. What choices did the trainee in Dr. Slutsky's laboratory have when she was told about the gift authorship? The paper was already published, thus any change of authorship would have involved admitting the situation to the journal. Asking your new boss to admit publicly to conferring gift authorship may not be a good way to begin your research career, nor is contacting the journal directly and triggering an investigation. Few would suggest driving a scientist from his faculty position for an isolated incident of gift authorship. Yet, some early action might avert subsequent, more serious problems.

Two points can be made here. First, there exists an anonymous procedure at most universities and companies for seeking advice if you encounter a situation such as gift authorship; typically, an officially designated ombudsman will help you resolve conflicts and difficult situations. Second, an adviser who puts you in a precarious situation is likely not the right adviser for you. The consequences of confronting a situation like this early are not likely as bad as they seem, while the consequences of avoiding confrontation are likely much worse.

Studies of scientific authorship often define guest authorship separately as listing a colleague who did not contribute directly to a paper in the hope that his or her reputation will enhance the odds of acceptance for publication (Davidoff, 2000). Combining gift and guest authorship into a single category, termed *honorary authorship*, Flanagin et al. (1998) surveyed the authors of papers in six major biomedical research journals (including *Annals of Internal Medicine*, *JAMA*, and the *New England Journal of Medicine*) and found that 19% of those publications had evidence of honorary authorship. They also found that 11% had evidence of ghost authorship, defined as the omission of an author who contributed significantly to the publication. Ghost authors may be junior colleagues who simply did not receive the credit they deserved, but they may also be professional medical writers hired to write articles anonymously or even representatives from companies with a financial interest in the findings who wish to hide their involvement.

A 1993 survey of postdoctoral research fellows at the University of California–San Francisco suggests even higher rates of inappropriate practices: 38% of the respondents thought that at least one coauthor on their papers was undeserving, while 20% thought they were excluded on at least one paper for which they deserved authorship (Eastwood et al., 1996). One of the most telling results of this particular survey was evidence that trainees who have unfavorable initial experiences with authorship lose faith or interest in the integrity of the system. Overall, 32% of the fellows surveyed said they would be willing to list an undeserving author on a paper if it would enhance the probability of publication or otherwise benefit their career; that number jumped to 72% among those who reported a previous adverse experience with authorship.

6.3.3   Financial Support

Accepted practice regarding authorship varies by field and by culture. While relatively rare, some strongly hierarchical departments expect that the chair of the department should be listed as an author (possibly even senior author) on every paper, regardless of contribution. Such an environment may pose a challenge for younger investigators who disagree with the policy, especially if following the expected procedure weakens their own publication records by preventing them from assuming senior authorship on their own work. It is important to recognize and discuss cultural variations when working in a group composed of colleagues who trained under different systems and when collaborating internationally.

6.3.4   Quid Pro Quo

Nearly everyone agrees that gift authorship is wrong, yet there are many related cases where a colleague who has contributed to a study in some way requests or expects authorship in return. The most common situation involves valuable resources such as antibodies or transgenic mice. Consider a situation where an investigator devotes significant time and energy developing such a resource, then publishes a paper describing it. Colleagues then ask for access to the resource for studies they wish to conduct. It is not uncommon for the investigator to offer to provide access in exchange for authorship on the resulting paper(s).

This basic situation has unlimited variations. At one extreme, a request for a resource can lead to a genuine collaboration on a new study that is reflected accurately in coauthored papers describing the results. At the other extreme, however, the situation can approach scientific extortion, with the original developer of the resource demanding authorship in exchange for access, knowing few colleagues will deny the request due to the substantial time and effort required to replicate the resource. While many who disagree with such arrangements accept them as a fact of life, some defend the practice, regarding authorship on related papers as appropriate reward for developing the resource.

Most believe that the appropriate reward for any innovation, whether a new equation, method, antibody, or transgenic mouse, is citation, not authorship. Colleagues who employ the innovation cite the original publication, giving appropriate credit to its originator. In the case of an equation or its solution, the original paper contains everything colleagues need. In the case of a transgenic mouse, the original paper contains only a description of how to generate such a mouse. Is it reasonable to expect the scientist who first generates the mouse to send mice to any colleague who requests them? Does the answer change if federal or state resources funded the original development of the mouse, as with most biomedical research? These and related questions about access to resources and data from publicly funded science are currently a topic of vigorous discussion in the scientific community; they are explored again in Chapter 8.

6.3.5   Students and Technicians

We have highlighted some common problems related to authorship beginning with the simplest and least controversial and proceeding to the more complex and controversial (and therefore interesting). Next, we consider the key question of who should or should not be an author on a particular paper or, to generalize the problem, of exactly what qualifies someone to be an author. Before proceeding, use the following exercise to define better what you think should be considered in making decisions about authorship.

Exercise 6.5   List up to five minimum criteria needed to justify authorship on a scientific or engineering paper. According to your criteria, would a laboratory technician or an undergraduate student who orders supplies and prepares samples qualify as an author on papers produced by the laboratory? What about a technician or student who runs tests according to instructions and turns over the data for analysis? What about one who runs tests, analyzes data, and makes a figure for the paper but does not write any of the text?

   1.
   2.
   3.
   4.
   5.

As this exercise illustrates, it is remarkably difficult to articulate general guidelines for authorship that provide practical guidance. Common responses to this exercise are that each author should make a "significant" contribution to the work, that each author should make an "essential" contribution to the work, or that each author should make an "intellectual" contribution to the work. This last point illustrates general agreement that a student or technician who simply prepares samples or collects data without a true understanding of the project should be acknowledged, not listed as an author. Nevertheless, none of these statements provides practical guidance.

To increase our appreciation of this situation, it is useful to consider the contribution of a potential author against the backdrop of what is required to produce a paper. First, one must generate an idea or identify a problem, then plan an approach to address the problem. Next, one must perform the study and collect the data or solve the equations. Analysis and interpretation of the data or results then precedes writing the paper, which typically requires a comparison to previous related findings. The example of a student or technician who only collects data or runs a computer code as instructed suggests that an author should be involved in more than one aspect of the study; if that person also analyzes data and summarizes the results for the paper, the claim to authorship would be stronger. Requiring involvement in multiple aspects of a study would limit the quid pro quo arrangements discussed above to cases where involvement went beyond providing a particular resource. It seems reasonable to stop short of requiring every author to participate in every phase, however. For example, most investigators would support authorship for a person who joined a group after the study was conceived and planned but otherwise was involved deeply in all aspects of a study.

The concept that all authors should be involved in multiple aspects of a study (e.g., design, experiment, analysis, interpretation, or writing) seems reasonable. Nevertheless, your list from Exercise 6.5 likely includes additional criteria. Must every author understand everything in the paper? Must every author read the final version before submission? Recalling that some of Dr. Slutsky's coauthors experienced the stigma of being authors on fraudulent papers, should every author review the original data that form the basis for the conclusions? Each investigator must wrestle with these questions over the course of a career; your answers to these questions may well evolve with experience. It is important to think carefully about these issues early in your career so that you can develop practices consistent with the ethical standards you set for yourself.

6.4 CURRENT STANDARDS AND EMERGING IDEAS

Many people have thought about ways to improve upon practices used to define authorship in the archival literature. In particular, some professional societies and journals have introduced simple practices that reflect more accurately the contributions of those involved in a publication. These practices also have the beneficial effect of forcing increased discussion among coauthors on issues related to authorship.

6.4.1 International Committee of Medical Journal Editors Standards

The International Committee of Medical Journal Editors (ICMJE) evolved from meetings that began in 1978 to establish guidelines for the format of manuscripts submitted to medical journals. This group regularly revises and disseminates the document "Uniform Requirements for Manuscripts Submitted to Biomedical Journals: Writing and Editing for Biomedical Publication," available at their Web site, www.ICMJE.org. This Web site also lists journals that adopted these standards. The uniform requirements continue to provide guidelines on style and format for articles in biomedical journals and also guidelines on ethical aspects of writing and reviewing journal articles.

The uniform guidelines provide criteria for deciding who should be an author on an archival paper. Although these guidelines are similar to principles discussed in the previous section, they have provoked objections from many scientists who believe them to be too strict (Bhopal et al., 1997); they are also rarely enforced, even by journals that claim to have adopted them (Davidoff, 2000). Under the ICMJE guidelines, all authors must meet the following three criteria:

1. Substantial contributions to conception and design OR acquisition of data OR analysis and interpretation of data.
2. Drafting the article OR revising it critically for important intellectual content.
3. Final approval of the version to be published.

In general, it is difficult to assess whether authors follow (or are even aware of) these criteria when submitting articles to journals that have formally adopted them. As discussed in Section 6.4.3, one study that attempted to test compliance with these guidelines found that only 56% of authors of articles in a prestigious journal that subscribes to the guidelines actually fulfilled them (Yank and Rennie, 1999).

Finally, as another example, the American Heart Association (AHA) publishes multiple outstanding scientific journals dealing with cardiovascular health and disease. Among other things, the AHA form entitled "Authorship Responsibility and Copyright Transfer Agreement" stipulates that to qualify for authorship, one must have participated in one or more of the following:

conceived and designed the research
acquired the data
analyzed and interpreted the data
performed statistical analyses
handled funding and supervision
drafted the manuscript
made critical revisions of the manuscript for important intellectual content

Other journals continue to require a simple statement that all authors contributed to the work and agree to its submission for consideration for publication (recall Section 3.1.1).

6.4.2   Author Notification

One of the simplest recent innovations is that many conferences and journals now require the submitting author to provide e-mail addresses for all authors, who are notified electronically of the submission of an abstract or manuscript. While no coauthor should ever learn of a submission for the first time through such an e-mail, this is not an infrequent occurrence. Notification allows an investigator who was unaware of a submission to raise objections while the abstract or manuscript is under review, rather than being forced into the much more difficult position of addressing the issue after acceptance or publication. Notification may also increase the odds that the submitting author will discuss the submission with all coauthors in advance to avoid surprising colleagues. Electronic notification is not a foolproof defense against those who are willing to forge the names of coauthors on a submission agreement. Those intending to deceive could easily construct false e-mail accounts for coauthors, but at least this would require more effort than simply forging a signature.

6.4.3 Specifying Contributions

A more radical approach is to discard the traditional premise that all authors bear equal responsibility for the content of an archival paper. Instead, some journals now ask authors to specify their contributions to an article at the time of submission. In theory, responsibility for integrity of the research partitions accordingly, with authors only responsible for ensuring the validity of their work. In addition, most journals require at least one author to declare responsibility for oversight of the entire article.

Specifying individual contributions simplifies attribution of responsibility or blame. It could also allow societies or journals to impose more uniform standards for authorship. For example, a journal could refuse authorship to anyone unwilling to take responsibility for more than one aspect of a publication. Partitioning responsibility may prove the only practical solution for large multi-investigator projects. Nevertheless, this approach changes the traditional understanding of an archival publication and meaning of authorship. It could have the disadvantage of weakening scientific collaborations, as papers increasingly become a compendium of individual miniprojects. Such a weakening is certainly contrary to what most of us envision when we discuss the need to foster more and better multidisciplinary collaboration on today's increasingly complex scientific and engineering problems.

One of the first journals to ask authors to specify their contributions as part of the submission process was the medical journal *The Lancet*, a signatory to the ICMJE Uniform Requirements for Manuscripts Submitted to Biomedical Journals discussed in Section 6.4.1. During the first 6 months after authors began specifying contributions, Yank and Rennie (1999) studied the reported contributions with three goals: to determine how author contributions related to position in the author list; to determine whether self-reported author contributions fulfilled the ICMJE guidelines; and to determine the degree of overlap between the contributions of those listed as authors and those listed in acknowledgments. They made the generous assumption that all authors read and approved the final version (ICMJE criterion 3), but they found that only 56% of authors fulfilled the other two criteria. Specifically, 78% of authors reported participating in conception, design, analysis, or interpretation (ICMJE criterion 1), a finding that was consistent for those who were listed first, second, third, or last in the author list. By contrast, 65% reported participating in writing or revising the paper (ICMJE criterion 2), with a range from 84% for the first author to 54% for the third author. The Yank and Rennie study contains other interesting findings, hence we recommend this study as a basis for a journal club or group discussion on authorship.

6.4.4   Quantifying Contributions

A natural response to uncertainty, especially among scientists and engineers, is to introduce quantitative measures. In addition to specifying what each author did, some have advocated specifying each author's percent contribution to the overall work. This is probably most common during tenure evaluation, when a junior professor under consideration for tenure estimates his/her percent contribution to each published paper. This is a difficult question to answer, especially during a tenure evaluation, because the desire to report strong contributions for yourself may tempt you to devalue the contributions by your coauthors. Typically, a group of collaborators who estimate the contribution of each group member produce percentages greater than 100% unless some mechanism (such as an interactive form or pie chart) constrains the total.

Another quantitative approach appeared in the biostatistics literature, reflecting the unique role played by many statisticians in research. Statisticians may be involved in design and analysis for many different studies but not directly involved in collecting data, performing experiments, or writing papers for any of those studies. This "specialist" role makes it difficult to apply typical criteria for authorship. As a possible solution to this problem, Parker and Berman (1998) proposed a scoring system to help decide when statisticians should or should not be listed as authors. Their system divides the statistician's role into three phases of a research project (design, implementation, and analysis) and requires for authorship either a deep involvement in two of the phases or a deep involvement in one and moderate in the other two. They also propose that it is unreasonable to hold a statistician who is listed as an author responsible for the integrity of the entire published article.

6.5   OUR APPROACH

As is common when discussing interesting ethical issues, we raised many more questions than we answered in this chapter. What is most important is that each person utilizes cases and questions such as those presented herein, as well as discussions with advisers and senior colleagues, to establish individual principles about authorship early in a career. *It is impossible to do what you think is right if you do not know what you think is right.* Once you establish your principles, the question remains of how best to put them into practice. In this section, we offer some of our own experiences as examples of how to apply a set of principles to the everyday practice of science and engineering.

6.5.1 Authorship Criteria

In our own groups, we expect that all authors on a paper should be involved in more than one aspect of a study, should agree to be listed as an author, and should be given a chance to contribute directly to the final version of the manuscript before submission. Ideally, each new group member and each new collaborator should discuss these criteria at the outset. At the very least, all members of the group working on a particular project must discuss issues of authorship before submitting the first abstract or publication related to that work. This is easiest to accomplish when all authors work at a single location and most difficult when the publication involves collaborators from different departments or institutions. Fortunately, the Internet and track-changes features in most word processing applications enable all coauthors to contribute directly to developing manuscripts regardless of physical location.

6.5.2 Predraft Group Meeting

In our experience, one of the simplest and most useful ideas is to convene a meeting of all potential authors to review findings and interpretations as well as to agree on authorship before writing an abstract or manuscript. The senior investigator who is funding or driving the project calls the meeting, inviting all contributors who potentially satisfy the criteria for authorship. In cases of coauthors from multiple locations, Web conferencing or teleconferencing becomes a vital resource. At the meeting, each contributor presents results to the group and answers questions. Then, the group discusses proposed figures, the proposed author list, and the choice of journal for submission. Notwithstanding the effort required to bring everyone together for an hour or two, this approach allows all potential authors to gain confidence in the validity of the studies, to ask questions and comment on the results and their importance, and to voice any concerns about the content of the paper, interpretation of the results, or author list before the bulk of the writing begins. This approach also helps improve the paper by subjecting the results to a round of "internal review," helps graduate students and fellows practice oral presentation skills, and helps strengthen relationships among collaborators.

6.5.3 Final Review and Approval

Once a manuscript has become a final draft, it is essential for all authors to review and approve the draft before submission. This is also an appropriate time to settle final questions of authorship, especially if no previous discussion has taken place. One reasonable approach is to list as authors on the draft those colleagues you believe merit authorship, but to include in the distribution list other people who have made some contribution and may feel they should be authors. Ask each recipient whether they feel they deserve to be an author on the paper (or whether they agree with the proposed author list) and whether they have any comments or suggestions for the manuscript before submission; follow up with those who do not reply. Like the predraft group meetings, this step ensures that all authors are aware of the content of any publication bearing their names and provides a round of internal review to improve the manuscript before external peer review.

Most investigators basically agree on the rules of authorship and are willing to follow them. Inappropriate attribution of authorship usually reflects someone succumbing to real or perceived external pressures or simply not giving the matter sufficient attention, rather than attempting to deceive. In general, our experience with regard to questions of authorship has been heartening. In most cases where a claim to authorship appeared marginal to us, our colleagues have responded to our question of whether they want to be an author, as we would have hoped, by stating that their contribution merits an acknowledgment rather than authorship. Many have provided helpful comments on a draft even after stating that they did not wish to be listed as authors. Perhaps surprisingly, our most difficult experiences have typically involved refusing authorship offered by a colleague rather than denying authorship to a colleague.

6.5.4   Default Position for Abstracts

The process described above is time-intensive. A confounding situation that can arise, therefore, is the last-minute abstract for consideration for presentation at a technical meeting. Such abstracts are short and typically have a fixed deadline for submission, thus they are often written just before the deadline. On such short notice, the collaborators involved in a particular study may not be available to meet to discuss the abstract or even to read, revise, and approve the final submission. In such cases, it is best to agree ahead of time on a "default" position for last-minute abstracts — if contributors cannot be reached to review an abstract on short notice, do they prefer to be listed as an author and review the abstract after submission or do they prefer to be left off the author list? We recommend the latter approach, for it is dangerous practice to include authors who have not read, revised, and approved the abstract before submission. We also note that it is appropriate to delay submission when a coauthor cannot be reached; there will always be other meetings and thus other opportunities.

CHAPTER 7

Recordkeeping

Scientists and engineers must keep records of their work, using a combination of laboratory notebooks, images, file folders, and electronic data. Similarly, clinicians must record each step of diagnosis and treatment in a patient's medical records. Although keeping precise records may seem mundane, those records are central to many important decisions in science, engineering, medicine, and public policy.

Exercise 7.1   Based on your experience in a research laboratory, or a laboratory course if you have not yet worked in a research laboratory, answer the following questions before continuing:
1. Did you maintain a laboratory notebook?
2. If yes, what instructions were you given about what to record?
3. If no, where did you record information related to the work?
4. Did your supervisor review your notebook or records? If yes, how often?
5. If someone tried to reconstruct your work from these records, what percentage could they reconstruct without your help?

If possible, compare your answers to those of a colleague who has worked in the pharmaceutical or medical device industry. It is likely that your answers will differ substantially; discuss the most likely reasons for this.

7.1   THE SLUTSKY CASE REVISITED

In Chapter 6, we considered the case of research fraud by Dr. Robert Slutsky, as described in a 1987 article in the *New England Journal of Medicine* (Engler et al., 1987), and we asked what aspects of this case were most surprising. In response to this question, many cite the following paragraph from the section entitled "What is Fraud?"

> After due consideration of what requirements and standards applied, the . . . committee adopted the position that the ethos of scientific research requires that hypotheses be validated before they can be accepted and that claims to observation be open to

scrutiny by peers. The legal principle of "innocent until proved guilty," which might be rephrased as "assume correct until proved wrong," does not apply to scientific work; the burden of proof remains with those claiming new findings. Thus, the authors of a scientific publication that is reasonably alleged to be fraudulent bear the responsibility for establishing the accuracy of their results.

Engler et al. (1987)

This excerpt should be sobering to anyone involved in research or development. Most of us have lost records to a computer crash, accidentally overwritten a file, lost a notebook, discarded old data, or at times kept less than complete records. If the supporting data are missing and the burden of proof against an allegation of fraud lies with the researcher, an anonymous accusation of fraud from a disgruntled colleague, employee, or student could be enough to support a finding of research fraud and end a career.

The proposition that the burden of proof lies with the researcher raises two questions. First, do you agree with the argument that the nature of science and engineering should place the burden of proof on the investigator, or should it rest (as in criminal law) with the accuser? Second, is the burden of proof actually placed on the researcher in current practice? While the first question provokes interesting discussions in any room of scientists or engineers, most are surprised to learn that the answer to the second question is a resounding *yes*.

The Office of Research Integrity (ORI) of the U.S. Department of Health and Human Services (http://ori.hhs.gov/) performs a range of functions designed to maintain the scientific integrity of biomedical and behavioral research funded by the U.S. Public Health Service (PHS). One of these functions is investigating and issuing reports on scientific fraud or misconduct involving PHS grants. Moreover, to heighten awareness of the importance of scientific integrity, the ORI publishes Findings of Scientific Misconduct (i.e., brief reports summarizing each case and its outcome) on their Web site and within weekly electronic mailings on funding opportunities distributed by the NIH. A review of past cases demonstrates that the burden of proof against an allegation of research fraud does indeed rest with the researcher. As an example, we reproduce below, in its entirety, a Finding of Scientific Misconduct issued in 2000. Particularly relevant to our discussion is that the ORI found that "Dr. Duan . . . engaged in scientific misconduct by reporting research that was inconsistent with original data or could not be supported because original data were not retained," even though "Dr. Duan denies all allegations of scientific misconduct and contends that some of his original data is missing."

Exercise 7.2   Read and discuss with a colleague the following Finding of Scientific Misconduct. What aspects of this report do you find surprising? What impact did the sanctions likely have on Dr.

Duan's career? Do you agree with the practice of publicly distributing these findings and naming the researcher involved? What impact might the fraud in this case have had on other researchers, doctors, or patients? Given the impact of the fraud, was the severity of the imposed sanctions appropriate?

FINDINGS OF SCIENTIFIC MISCONDUCT
Release Date: June 27, 2000
NOTICE: OD-00-043 Department of Health and Human Services

Notice is hereby given that based on oversight by the Office of Research Integrity (ORI) and decision by the Assistant Secretary for Health, the U.S. Public Health Service has taken final action in the following case: Lingxun Duan, M.D., Thomas Jefferson University: The U.S. Public Health Service (PHS) alleges that Dr. Duan, former Research Assistant Professor of Medicine, Division of Infectious Diseases, Department of Medicine, Jefferson Medical College, Thomas Jefferson University, engaged in scientific misconduct by reporting research that was inconsistent with original data or could not be supported because original data were not retained.

The research in question was supported by a National Institute of Allergy and Infectious Diseases (NIAID), National Institutes of Health (NIH), grant, R01 AI36552, entitled "Intracellular antibodies and HIV 1." Specifically, the research in question was reported in an NIAID, NIH, grant application; in an FDA-approved phase I gene therapy investigational new drug (IND) application entitled "Intracellular immunization against HIV-1 infection using an anti-rev single chain variable fragment (SFV);" and in two publications: (1) Duan, L., Bagasra, O., Laughlin, M.A., Oakes, J.W., & Pomerantz, R.J., Potent inhibition of human immunodeficiency virus type I replication by an intracellular anti-Rev single chain antibody, Proc. Natl. Acad. Sci. USA 91:5075–5079, 1994; and (2) Levy-Mintz, P., Duan, L., Zhang, H., Hu, B., Dornadula, G., Zhu, M., Kulkosky, J., Bizub-Bender, D., Skalka, A.M., and Pomerantz, R.J., Intracellular expression of single-chain variable fragments to inhibit early stages of the viral life cycle by targeting human immunodeficiency virus type 1 integrase, J. Virol. 70:8821–8823, 1996.

Dr. Duan denies all allegations of scientific misconduct and contends that some of his original data is missing. Both Dr. Duan and PHS are desirous of concluding this matter without further expense of time and other resources. Thus, Dr. Duan has entered into a Voluntary Exclusion Agreement (Agreement) with PHS, in which Dr. Duan has voluntarily agreed:

(1) to exclude himself from any contracting or subcontracting with any agency of the United States government and from eligibility for, or involvement in, nonprocurement transactions (e.g., grants and cooperative agreements) of the United States Government as defined in 45 C.F.R. Part 76 for a period of two (2) years, beginning on June 7, 2000;

(2) that for a period of one (1) year after the conclusion of the voluntary exclusion period, any institution that submits an application for PHS support for a research project on which his participation is proposed or that uses him in any capacity on PHS supported research, or that submits a report of PHS funded research in which Dr. Duan is involved, must concurrently submit a plan for supervision of his duties to the funding agency for approval; the supervisory plan must be designed to ensure the scientific integrity of Dr. Duan's research contribution, and the institution must also submit a copy of the supervisory plan to ORI;

(3) to exclude himself from serving in any advisory capacity to PHS, including, but not limited to, service on any PHS advisory committee, board, and/or peer review committee, or as a consultant for a period of two (2) years, beginning on June 7, 2000;

(4) that he will not oppose the submission to journals of a statement summarizing the current state of the science with respect to the scientific matters at issue relating to grant R01 AI36552, which has been jointly agreed to by Thomas Jefferson University and the United States of America.

FOR FURTHER INFORMATION CONTACT: Acting Director, Division of Investigative Oversight Office of Research Integrity 5515 Security Lane, Suite 700 Rockville, MD 20852 (301) 443-5330

7.2   WHY KEEP RECORDS?

Accurate records are central to any investigation of scientific misconduct, yet such investigations are rare. Not surprisingly then, defense against an accusation of misconduct is not the primary reason researchers keep records, and this potential concern should not dominate our discussion of record-keeping. A discussion of what records to keep and how best to do so begins with a consideration of what information will be needed in the future and why.

Exercise 7.3   First, list reasons why physicians write information in a medical chart. Compare your list with one or more colleagues and add to your list as needed until you believe it is complete. Second, make a similar list of reasons that researchers at a medical device company record information in laboratory notebooks. Compare this list with your list for medical charts; how many of the reasons for keeping records appear on both lists? Third, list reasons why a researcher working in academia records research methods or findings. Are there any reasons unique to this third list?

7.2.1   Medical Records

Although your list may differ, commonly cited reasons for writing in a medical chart are immediate transfer of information, long-term transfer of information, training medical students and residents, and legal documentation. Examples of immediate information transfer include a physician writing an order in a chart that another member of the hospital staff must execute later in the day, or a resident who is called in the middle of the night to examine a patient deciding an appropriate course of action based in part on his or her review of the patient's chart. Because many different people come in contact with each patient during a typical day in a hospital, a smooth transfer of information can literally be the difference between life and death.

Availability of an accurate longer-term medical history can be equally important to a patient's health. Diagnosing and treating a patient often depends critically on details of that person's medical history: previous illnesses and surgeries, current medical problems and medications, allergies, and so forth. Few patients will remember, or even know, all the details of their own medical history, and few physicians can remember the complete histories of patients under their care. Consequently, a written record of each patient's medical history is not only essential to the accurate exchange of information between physicians, it is also critical as an accurate, detailed substitute for each physician's memory.

Perhaps less obvious, good recordkeeping can be useful in training medical students, nurses, and other health care professionals. Most entries in medical records have very specific formats. Learning and using these specific formats is integral to learning the thought process associated with medicine. One usually records a detailed medical history and results from a physical examination on a form that lists standard questions and aspects of the examination. Recording the same information for each patient helps students learn the essential components of a good examination; they soon begin to ask the questions and perform the examination in the same order each time, which helps ensure that they do not miss anything. Another common entry in hospital charts is the SOAP note, an acronym for "subjective, objective, assessment, and plan." Organizing daily updates under these four headings encourages a particular thought process: gather the information, think about what it means, then decide what to do.

Finally, it is no surprise that medical charts serve as an important legal record of what happened to a particular patient and why. In fact, most respondents to Exercise 7.3 place this first in their list. Unfortunately, many increasingly view this legal function as conflicting with the training function discussed above. Many hospitals no longer allow medical students to write in a patient's chart for fear that an erroneous assessment or plan, even if corrected later by the supervising physician, could increase vulnerability in a lawsuit.

7.2.2   Industry Research Records

Your second list from Exercise 7.3 may not differ much from the first list. Laboratory notebooks kept by employees of a medical device company serve many of the same functions as a medical chart. Multiple technicians might record results from a series of tests for review and compilation by their supervisor the next day (short-term transfer of information); technicians may consult their records when performing the same tests a month later to make sure they set them up exactly the same way (long-term transfer of information). Asking new employees to follow a specific structure for recording data from a particular test can help them learn how to perform that test (training). Finally, approval of new drugs or devices by the U.S. Food and Drug Administration requires stringent recordkeeping (legal documentation); such records are thus essential for the survival of pharmaceutical and device companies.

If you or a colleague with whom you discussed Exercise 7.3 has worked in industry, the topic of cosigning likely surfaced in your list. Most industrial research facilities require a supervisor to review and cosign laboratory notebooks at the end of each day. This requirement can assist many of the functions of recordkeeping discussed above. If a test result is surprising, a supervisor can learn about the result and take appropriate action immediately: check the equipment, schedule a repeat test, or discuss the findings with his or her boss (information transfer, quality control). Daily review also provides an excellent opportunity for feedback on how best to perform the test or record the results. Finally, for companies that depend critically on regulatory approval of their products, cosigning not only helps ensure proper performance, it also properly documents all tests and procedures (legal).

7.2.3   Academic Research Records

By now, the pattern should be apparent; recordkeeping serves similar functions in diverse disciplines and settings. Academic researchers use records to transfer information between members of a group, as a long-term record of what was done and how it was done, to help train students to perform and record their work, and as a legal record. Because most academic researchers are more interested in publishing journal articles than protecting themselves against product liability suits, the need for long-term documentation tends to dominate in academic practice. Nevertheless, answers to Exercise 7.1 usually reveal that many research groups do not keep adequate records, even to meet the basic goal of documenting what research was performed, when, and by whom. One of many interesting findings that emerged from a 1993 survey of postdoctoral research fellows at the University of California–San Francisco was that fellows with an M.D. degree were significantly more likely to keep laboratory records in ink in a permanently bound research notebook than were those with a Ph.D. degree (Eastwood et al., 1996).

Exercise 7.4   Design a recordkeeping policy for your research group. What should be recorded and where? Should cosignatures be required? If yes, who should cosign and how often? Should cosigning or other rules of recordkeeping differ for different members of the group (e.g., undergraduates, graduate students, postdoctoral fellows)? How should new members who join the group be instructed in keeping records? Who should be responsible for ensuring that rules are followed? What should be the consequences for a group member who fails to keep appropriate records? What should happen to a member's records when they graduate or leave the group? Finally, are any special rules needed for electronic data? How does your policy compare to your own current practices in your research?

7.3   ELECTRONIC DATA

Consider a spreadsheet or data file containing results from a dozen experiments performed over a 6-month period. If you had produced this data file and were asked to verify that you actually performed the experiments, what proof could you offer? The data file itself is of little use; one can numerically generate data and save the results in the desired format. The operating system will likely display the date you created the file and the date you last modified it, but these dates can be manipulated by changing the computer's clock. If you had kept every version of the file as you entered data from each new experiment, paging through these versions would be more convincing, but this trail of different versions of the file could also be fabricated. Having different versions of the file plus a laboratory notebook showing data acquisition on dates that match those for the files would be better, especially if the notebook had been cosigned by a supervisor at intervals over the period in question.

That most research today relies either entirely or in large part on storing results electronically presents enormous challenges for ensuring integrity of the science and engineering. While researchers rarely need to prove that they performed specific experiments, they often need to revert to an older version of software or refer to computer files that have been deleted or altered. Consider this fairly common dilemma from the perspective of the director of a research laboratory at a university. A new student begins working for you, taking over a project from a previous student who graduated. Her project involves imaging cells and analyzing the images with custom-written software that is still under development in your group, and her early results appear to contradict findings by her predecessor. To differentiate between possible changes in experimental protocol, methods of collecting the images, misusing the current version of the software for analysis, and consequences due to recent changes in the software, you will need a trail of previous images, versions of the software, and well-annotated analyses. Few laboratories proactively anticipate such situations in their recordkeeping or established procedures for data backup, but most will experience them.

7.3.1   Date-Stamps, Time-Stamps, and Backup Systems

The most rudimentary form of information about a digital file is its date- and time-stamp. These stamps have little weight in an investigation of potential fraud because of their ease of manipulation, but they can be very helpful in typical situations wherein everyone acts in good faith. In the example just discussed, trying to reconstruct the version of a particular program used for a particular analysis could be aided greatly by comparing the date-stamp of the file containing the results with dates associated with successive versions of the software. The major problem with date- and time-stamps is that they can be changed easily, even inadvertently. Simply opening, printing, or resaving an old file may result in a new date- and time-stamp, depending on the setup for autosaves. A cumbersome solution is to lock each data file when creating it, thus forcing subsequent changes to be made on copies of the original. A more common approach is to create periodic tape or hard-drive backups of data files. This approach stores both the current versions of the files and their current date- and time-stamps, thus enabling one to recover deleted or altered information and to reconstruct the history of modifications to a particular file. Given the low cost of disk space, systematic routine backup of all electronic data is a prudent and affordable safeguard.

7.3.2   Images

Digital images raise many of the same questions as other electronic data files. Because image files are often large, there is a temptation to delete an image following analysis, thus retaining only final results that occupy less storage space. In general, however, it is more effective to do exactly the opposite: store copies of the original images on recordable media such as CD-R and force group members to manipulate only copies of the original image. Indeed, in cases where an analysis subsequently becomes suspect, one can always return to the original image and reinterpret the findings.

Most medical images contain information that identifies the patient, which raises another important concern with regard to image handling. Recent efforts to protect a patient's privacy have led to important new regulations and training for everyone who views, analyzes, or handles medical images. Similarly, concerns about potential attacks on researchers and facilities has led many universities to formulate policies for handling and storing images that show research with animals. Given the imperfect nature of computer security, many policies prohibit the storage of sensitive images involving patients or animals on any computer connected to the Internet.

7.3.3   Software Development

When seeking solutions to a problem, it is often useful to consider who has the best incentive to solve that problem; they will often have the best solution. Developing software often requires many people to work on different aspects of a single code, integrating changes made by different group members in an orderly way while tracking previous versions so that problematic changes can be undone. It is not surprising, therefore, that the software industry has developed excellent systems for controlling revisions. A number of these systems are now available as open-source software, making such solutions accessible even to research groups whose main business is not software development.

Exercise 7.5 Now that we have discussed electronic data in more detail, return to the recordkeeping policy you designed for your research group in Exercise 7.4. Provide more specific instructions for the primary types of electronic data in use within your group. What would your group need to change to implement your plan? If you would need additional servers or individual external hard drives for routine backups, estimate the amount of storage space you would need and determine how much this would cost. If you believe your group should implement software revision control, identify open-source and commercial products that might meet your needs. If you feel strongly that changes are needed, consider discussing potential changes and options with your colleagues at a group meeting.

7.4 FRAUD: FABRICATION AND FALSIFICATION

Even in a book dealing with the ethics of communication in science and engineering, it seems futile to admonish readers not to falsify or fabricate data. Very few would willingly do so, and the remaining few will not be stopped by our disapproval. We focus, therefore, on the need to clarify the boundary between ethical and unethical behavior, not on why one should behave ethically — we assume that the latter is clear.

Exercise 7.6 Most universities provide extensive information related to issues of academic dishonesty and misconduct. For example, undergraduate students at Texas A&M University are referred to http://ugr.tamu.edu/resources/. Search three university Web sites and record their definitions of *academic dishonesty* as well as specific actions that constitute such dishonesty. Compare results and submit a three-page summary.

Before continuing, consider some widely accepted definitions related to academic or research misconduct, which in this case are found on the aforementioned Web site at Texas A&M University: Misconduct in research or scholarship includes fabrication, falsification, or plagiarism in proposing, performing, reviewing or reporting research. It does not include honest error or honest differences in interpretations of data.

Fabrication: Making up data or results and recording or reporting them.
Falsification: Manipulating materials, equipment, or processes, or changing or omitting data or results such that the findings are not represented accurately in the research record.

Plagiarism: The appropriation of another person's ideas, processes, results, or words without giving appropriate credit.

Recall, too, the following definition:

Fraud: A deception deliberately practiced in order to secure unfair or unlawful gain; a misrepresentation of material fact consisting of a false representation, concealment, or nondisclosure.

7.4.1 Retaining or Discarding Data

Consider a student who has carefully collected and plotted a dozen data points on an x–y graph, then fitted a line through the data. Fabricating extra data that lie near the best-fit line would improve the r^2 value typically reported to indicate how well the line fits the data; classifying a few points far from the best-fit line as outliers and discarding them would similarly improve the reported fit. Why is it that every scientist or engineer would consider fabricating extra data points to be fraudulent, while many would at least consider discarding outliers?

The *Nature* article titled "Scientists Behaving Badly" reported findings from a survey of more than 3000 NIH-funded scientists regarding the self-reported frequency of behaviors identified as concerning by focus groups of researchers (Martinson et al., 2005). While only 0.3% admitted to falsifying research data within the previous 3 years, 15.3% admitted to "dropping observations or data points from analyses based on a gut feeling that they were inaccurate." The only more frequent offense (27.5%) was "inadequate record keeping related to research projects."

Unfortunately, the high incidence of self-reported manipulation of data in the 2005 report in *Nature* agrees well with earlier surveys that asked research trainees about their willingness to commit various types of research fraud. In a 1993 survey of postdoctoral fellows at the University of California–San Francisco, 12% of respondents reported first-hand knowledge of a scientist intentionally altering data for a presentation, while 4% reported first-hand knowledge of data fabrication. The numbers were slightly lower, but substantial, for grant applications (8% alteration, 2.5% fabrication) and publications (8% alteration, 2.5% fabrication). Despite the respondents constituting a self-selected group of fellows interested enough in research ethics to return the survey, 15.4% indicated they would be willing to select or omit data to improve their results "if it would make publication of [their] work more likely or benefit [their] career," rising to 27.2% "if it would increase the chances of [their] grant application being funded."

Exercise 7.7   Formulate criteria for appropriately discarding a data point, observation, or study. Compare your draft policy with that of one or more colleagues and revise until you believe that your policy could be implemented in your group. Next, find a data set collected in your group that you know contains noise or errors and test your policy by applying it to the data set. How many points would you discard, if any, and what justification would you give? As a final step, discuss your analysis with colleagues, then with a senior colleague or mentor. Submit a three-page summary of your conclusions.

An Internet search will return many different criteria and algorithms for identifying "outliers" in a data set, yet such statistical analyses are rarely sufficient for excluding data. Exclusions must combine good experimental and statistical practices with an understanding of the overall study and potential impact of those exclusions. In general, exclusion criteria should be established before performing the study and should depend on something other than the values obtained. Excluding data points collected on a single day that appear to be outliers, based on the rationalization that the equipment must have been miscalibrated, is a dangerous practice. In contrast, performing a calibration at the end of each day and excluding from analysis all measurements taken on days where the calibration fell outside a preset tolerance is good technique. Discussing criteria for excluding data before performing a study can help ensure that you have enough information to make good decisions later. Other good practices include disclosing exclusions and the associated justifications within publications, discussing exclusions with your mentor before making them, and remembering that "if something does not feel right it probably is not."

7.4.2 Image Manipulation

The importance of digital images in many types of research raises additional questions about the degree of processing or manipulation that is appropriate in a given situation. Consider a gel such as a Western blot, where a series of stained bands indicate relative amounts of certain proteins (rows) in different samples (columns or lanes). Is there anything unethical about cutting and pasting from several images of gels run on different days to create a composite image showing results from certain samples side by side? What if one or more of those gels was underexposed? The resulting image would then be lighter than expected and this problem could be corrected in one of two ways: reimage the gel with a longer exposure time or adjust the image contrast and brightness using an image processing program. Are these two options really the same, or is manipulating the image fraudulent?

It turns out that your answer to this question may change if you learn more about how image processing programs adjust contrast and brightness and how densitometry programs quantify bands on a gel. Commercial densitometers often output optical density, which relates directly to concentration. Other devices (CCD cameras, scanners) commonly used to image gels produce images in which pixel intensity has a nonlinear (logarithmic) relationship to optical density; in such cases, manipulating images before analysis may have important consequences. When deciding how to interpret findings, it is not enough to want to make good decisions; it is critical to gather enough information to make good decisions.

7.4.3   Statistical and Image Forensics

After reading case summaries on the Web site of the ORI (http://ori.hhs.gov/misconduct/cases/), it is easy to become concerned. There are cases of undergraduates, graduate students, postdoctoral fellows, and principal investigators fabricating data, scientists manipulating images in applications for funding, and even a laboratory member altering another's experiments to ensure that attempts to repeat his earlier (fraudulent) experiments would not expose him. As you move forward in your career, you will begin to get more of your data secondhand from employees or students whom you supervise. You will also spend more time reviewing technical reports, manuscripts, or grant applications from other researchers. At some point, it is natural to wonder how best to check the validity of the data and images that you encounter.

Within your own group, good training, good recordkeeping, and replication of randomly selected experiments remain important. Even in an environment that emphasizes integrity, replication of selected experiments can help guard against error and ensure continuity of methods as members join or leave the group. When you have less information about the source of data in a figure or table, however, or when you have reasons to suspect that data or images have been altered, emerging tools for statistical and image forensics may help. The ORI Web site is a good resource for such tools.

Statistical forensics generally relies on the observation that those who fabricate data rarely do so with much statistical sophistication. Dr. Slutsky (discussed in Chapter 6) published two papers that contained data sets having different sample sizes but identical means and standard deviations. Other cases have identified fabrication via nonrandom distributions of the rightmost digit in a series of numbers, including a case wherein the rightmost digit was either 5 or 0, thus suggesting that the sample size had been inflated by averaging pairs of samples to fabricate points for additional, nonexistent samples.

Image forensics relies on software tools that detect the manipulation of images. One frequently used set of tools is called Forensic Droplets; it is available on the ORI Web site (http://ori.hhs.gov/tools/droplets.shtml). These droplets run in Photoshop and help the user detect common methods of manipulation. The Web site also provides illustrative uses of these droplets in actual cases of misconduct (http://ori.hhs.gov/tools/principles.shtml).

Exercise 7.8   Download and read the article, "What's in a Picture? The temptation of image manipulation" [Rossner M, Yamada KM (2000) *J Cell Biol* 166: 11–15]. This article summarizes conventions adopted by several top journals regarding accepted types of image manipulation and how they should be identified. This article is also interesting because it presents several images that were altered intentionally. Download the appropriate Forensic Droplets from the ORI (http://ori.hhs.gov/tools/droplets.shtml) and use them to see whether you can detect the manipulation of the images presented in Rossner and Yamada's paper.

CHAPTER 8

Ownership of Ideas, Data, and Publications

Imagine sitting down 20 years into a successful career to write a book on your area of expertise. Everyone would agree that it would be wrong to copy verbatim a paragraph from another scientist's paper without attribution; we call that plagiarism. On the other hand, most would find it reasonable to include in your book a figure from one of your earlier journal papers. You may be surprised that to do so would be illegal in most cases. The publisher of the journal probably holds the copyright on that figure, and either you or the publisher of your book must secure permission from the journal to reprint the figure; such permission may involve paying a substantial fee.

One of the many interesting questions in science is who owns the results. Recent trends, such as the drive to commercialize products of university research and to increase public access to data from federally funded research, highlight this question of ownership. Debates about ownership of and access to the results of scientific and engineering research have important consequences for individuals, universities, companies, publishers, the government, and the public at large.

In Chapter 7, we compared medical charts to laboratory notebooks when considering why we should keep records. This comparison also provides an interesting entry into the question of ownership of information. Who owns your medical record and what conceptual tests for ownership does your answer suggest? The information is about you and you can request a copy of the records, which suggests some level of ownership. Nevertheless, you typically cannot alter or destroy the original medical record, which suggests that someone else shares ownership. The physicians, nurses, and other medical personnel who produced the record can write in it and read it, but they cannot obtain a personal copy to take home. What about the insurance company who paid for your care? Does paying for the associated diagnostic tests and office visits give the company any stake in ownership of the information?

Exercise 8.1 Based on your experience in research and the analogy to a medical chart outlined above, discuss why each of the following parties should or should not be considered owners of scientific data (not the resulting publications or patents) that result from a research project funded by the federal government using tax dollars and performed at a private university:

1. The principal investigator of the study
2. The students, technicians, and fellows who performed the experiments
3. The university
4. The federal government
5. The public/taxpayers

Now, for each of the people, groups, or institutions listed above, specify appropriate levels of access to the original data. In other words, who should be able to acquire copies of all the data, to alter those data, or to use them to write papers or submit patents? Are there people who should not be considered co-owners but who should still be granted some access? Explain.

8.1 DATA AND RESOURCE SHARING

In 1999, Congress amended the FY 1999 Omnibus Spending Bill to require federal agencies that fund research to ensure that all resulting data be made available to the public under provisions of the Freedom of Information Act. This new requirement prompted an outcry from the scientific community (Frankel, 1999). Concerns ranged from how researchers could disclose data from clinical studies without violating the privacy of human subjects to whether colleagues, companies, or even political activists might use data obtained under the new policy to compete with or disrupt the work of individual scientists. Such concerns are not unique to biomedical science, but because the NIH funds so much science in the United States, we focus below on NIH policies regarding the sharing of data, model organisms, and publications resulting from federally funded research.

8.1.1 Research Data

Ultimately, the NIH adopted limited requirements for sharing data. The Final Statement on Sharing Research Data reads, in part, as:

> NIH reaffirms its support for the concept of data sharing. We believe that data sharing is essential for expedited translation of research results into knowledge, products, and procedures to improve human health. The NIH endorses the sharing of final research data to support these and other important scientific goals. The NIH expects and supports the timely release and sharing of final research data from NIH-supported studies for use by other researchers.

See http://grants.nih.gov/grants/policy/data_sharing/index.htm for more on the NIH statement. Note, however, that provisions outlined later in the statement exempt most researchers from this policy. Most importantly, only applicants requesting over $500,000 of direct costs in any year must file a data sharing plan. Because the most common type of NIH grant (the R01, which is discussed in Chapter 4) typically has annual direct costs of $250,000 or less, this provision exempts most NIH-funded researchers. In addition, the NIH defines "timely release and sharing" to be "no later than the acceptance for publication of the main findings from the final data set." It is not uncommon for a successful researcher to renew the same grant repeatedly over 20 or more years while studying a particular disease; in such situations, it is unclear what constitutes "the final data set." Other critical issues, such as how long an investigator must continue to share data after completing a study, are not addressed by the NIH policy. Finally, note that the wording refers to sharing of data for use by other researchers, possibly circumventing scientists' original objections to sharing data with companies and the general public.

Community practice in some fields has overtaken the debate on data sharing. For example, researchers studying gene expression using DNA microarrays have established open databases and standards for submitting data (http://www.mged.org/). Top journals in this field also typically require authors to deposit their microarray data in a database as a condition of publication. Interestingly, although arguments for data sharing typically focus on benefits to the scientific community or public, Piwowar et al. (2007) recently reported a direct benefit to researchers: papers associated with publicly available microarray data are cited more frequently.

8.1.2   Model Organisms

In contrast to its data sharing policy, the NIH's policy on model organisms is clear, demanding, and relatively uncontroversial. All grant applicants who plan to develop a model organism such as a transgenic mouse must provide a plan for sharing that model organism with other researchers. Peer reviewers evaluate sharing plans as part of the grant review process and NIH staff may require "adequate progress in model organism sharing as well as a demonstrated willingness to make research resources developed during the project widely available to the research community . . ." to continue funding an existing grant (see http://grants.nih.gov/grants/policy/model_organism/index.htm).

8.1.3   Other Research Products

Any researcher can quickly and easily share data in a computer spreadsheet by posting the file to a public Web site. Even those who do not wish to maintain a Web site can usually deposit tables of supplemental data with a journal at the time of publication. Annotating the data with clear headings, comments, and notes about exclusions, then answering occasional questions from colleagues about the posted data, requires a little more time, but not much. By contrast, maintaining a colony of transgenic mice and providing breeding pairs to any interested colleague can be time consuming and expensive. Should we require researchers to continue to breed and supply mice to colleagues even after the NIH grant that funded development of that mouse expires? What alternatives are available to the scientist who wishes to share a transgenic mouse without the expense of maintaining a mouse colony indefinitely?

Sharing of computational models and custom software presents similar difficulties. Developers of computational models typically publish a description of the model and its results, not the actual code. As with transgenic mice, another researcher who attempts to reproduce the model from the published description could easily invest significant time and effort and still generate a slightly different model. Requiring researchers to post raw code for models developed with public funding would help, but this raises questions about how much support the developer should provide to those who download and attempt to use the code and for how long. Software companies have a financial incentive to provide user-friendly interfaces and technical support, but individual researchers do not. Research-grade models and software may be difficult for someone other than the developer to understand and operate, but as long as the software performs its intended task, the developer may have little interest in making it more user-friendly or in writing a supporting manual.

The incomplete nature of published descriptions of models, uneven practices regarding sharing, and the variety of operating systems and programming languages worldwide all combine to limit the effective sharing of computational models. Fortunately, several organizations are focused on improved sharing and interfacing of computational models across groups. One such effort is the Physiome Project (http://www.physiome.org/), organized by the International Union of Physiological Sciences. One component of this effort is to use markup languages to standardize coding of models and handshaking between them. A visit to the CellML Web site (http://www.cellml .org/models) provides more information on this effort as well as an idea of the difficulty of implementing, documenting, and debugging computational models of biological systems based only on their published descriptions.

8.2   COPYRIGHT

We noted in the introduction to this chapter that technical journals typically hold the copyright to any articles they publish. As an author, you transfer copyright to the journal as a condition of publication, even if the journal assesses page charges and labels your paper an advertisement.

Exercise 8.2   You are writing a review article and wish to include several figures from your earlier papers. Most publishers of those articles agree that you can use the figures as long as you state that they are used with permission of the publisher and include the appropriate citation. One publisher demands several thousand dollars in fees to reprint your figures, however. List all the options you can think of, including paying the fees, and discuss the relative merits and ethics of each approach. Based on that discussion, what would you do?

8.2.1   Online Publishing

The past few years have brought forth the most dramatic shift in the history of scientific publishing since the printing press. Like newspapers, scientific journals have experienced a drop in individual print subscriptions and a shift to online delivery of content. The critical question for journals, as for newspapers, is how to remain solvent selling content to an online audience that is used to getting information free. Not surprisingly, different journals have taken different approaches. Most publishers continue to sell print copies of their journals to libraries, at least for now; some restrict online access to print subscribers to encourage print subscriptions; others offer online-only access at a lower price than print subscriptions; still others provide full, free access to anyone. Regardless, most journals partially defray costs of publication by assessing substantial fees for publication. Increasingly, publishers and third-party services sell universities and other large institutions bundled access to groups of journals. Students and faculty at a typical research university can now access most major scientific and engineering journals through the university library's portfolio of electronic subscriptions. These subscriptions often include extensive collections of scanned back issues, virtually obviating the need to visit the library when studying the literature.

One of the consequences of the shift of journal content to electronic format is that it is often unclear to the typical academic user in the United States which journals charge for content and which distribute it freely. If you access a paper through your library's electronic journals portal, it is equally easy to obtain a PDF of an article whether or not the journal normally charges an individual for the download. If you access the same paper from outside your library's portal, however, the download fees are often $20 to $30 per article. This charge explains why you, as an author, are likely to receive occasional e-mails requesting a PDF reprint of one of your articles; the request usually comes from someone who does not have free access to the article. As an author, should you send a PDF when you receive such an e-mail? Sending the requested PDF deprives the journal of the fee it would normally charge for a download; on the other hand, what if the request comes from a colleague in a country where the download fee represents an exorbitant sum? What about the fairly common practice of posting PDFs of recently published articles on your laboratory or group Web site? Is this unethical? Illegal?

8.2.2   Public Access to NIH-Funded Journal Articles

If the NIH funds your research, recent policy changes rendered moot the problem of whether to share PDFs of your article. Nevertheless, the NIH may have also placed you in an uncomfortable position between the people who fund your studies and those (often your main professional societies) who publish them. The NIH began by asking grantees to deposit a copy of any accepted manuscript resulting from NIH funding into a publicly accessible database, PubMed Central. The policy was initially voluntary, although many NIH investigators felt pressured to deposit manuscripts because they thought that applications for renewing grants would be judged in part on productivity and that the number of deposited papers would be used as a measure of productivity.

Many journals considered the NIH database a violation of their copyright on the accepted articles, and some that depended heavily on selling access to their content claimed the NIH requirement would put them out of business. The American Physiological Society (APS), which publishes a portfolio of journals on physiology, genomics, and physiology education, was one of the most vocal critics. Among other actions, the chair of the APS Publications Committee sent the following e-mail in November 2005:

Dear APS Author,

On February 3, 2005, the NIH announced a new policy (NOT-OD-05-022) to enhance public access to publications resulting from NIH-funded research. The policy itself and information about it are available at http://www.nih.gov/about/publicaccess/index.htm. As the Publications Committee Chair of the American Physiological Society (APS), I am writing because some confusion has arisen about the NIH Policy.

The NIH Public Access Policy is a voluntary program that applies to NIH-funded investigators. Such authors are asked to submit electronic copies of articles accepted by peer-reviewed journals that report research funded by the NIH to the National Library of Medicine's PubMed Central (PMC). The NIH will then make the journal-accepted manuscript free to the public at an interval (ranging from immediately to 12 months) after publication that is chosen by the author. This Policy goes into effect on May 2, 2005.

Since the policy was announced, questions have arisen about whether or not participation is truly voluntary. On the day the policy was published, NIH Director Elias Zerhouni sent a letter to all extramural scientists and their research institutions describing the policy and urging them to participate. Although Dr. Zerhouni stated that the policy is a request, many researchers, university officials, and even some NIH program officers have interpreted it as a mandate for grantees. However, in public statements, Zerhouni and other NIH officials have repeatedly underscored that it is voluntary and there will be no repercussions for those who choose not to participate. Funded investigators can still fulfill their progress report requirements by providing print copies of their publications with their annual progress reports.

While the APS does not support the NIH Plan, we do recognize that it does put you, our authors, in a difficult position. Do you abide by a request issued by the granting agency or do you abide by the copyright statement that you signed when you submitted your manuscript to the journal? The APS does not want to see you placed in that position. Therefore, we are modifying our copyright statement to help you fulfill the voluntary request of the NIH Plan.

In doing so, we ask that you recognize that the Society has been at the forefront of online publishing, putting content online as early as 1994, providing authors with one of the first online manuscript submission and review systems, and underwriting the scanning of APS journal content back to 1898. We were one of the first publishers to change our access policies so that all content is free to all 12 months after publication. These efforts have cost the Society millions of dollars and subscriptions are one of the few ways available for us to recover those costs.

With over 50% of articles published in our journals funded by NIH, free release of manuscripts by PMC sooner than the Society's access policies allow could lead to losses of subscription revenues that would interfere with the journals' ability to meet the needs of the Society and its members. Moreover, NIH is seen as a leader among biomedical funding agencies. If others including NSF, NASA, or funding agencies in other countries such as the Wellcome Trust follow suit, we may end up in a situation where the vast majority of content is subject to mandates requiring public release before the journal release date. Should this occur, the APS and other scholarly publishers may be forced to increase author fees to compensate. Ultimately it would be detrimental to science if the APS had to charge authors the full cost of publication, which is currently about $3,000 an article.

Given the importance of subscription revenue to the Society's ability to provide our members with high quality and innovative publications, the APS asks that if you choose to deposit your manuscript into PMC, you will specify that it should not be made available to the public until 12 months after publication in the Society's journals. The Society intends to modify its copyright agreement so that NIH-funded authors are granted permission to deposit their accepted manuscript into PMC for release to the public 12 months after publication. By abiding by the Society's modified copyright agreement, you will be able to participate in the NIH public access program while still protecting the ability of the APS to recover the costs associated with its publication program.

Thank you for your past and future support of the Society's journals. We will be able to continue to publish these respected journals with your recognition that the NIH Public Access Plan is a voluntary plan that seeks release at 12 months, a time consistent with the Society's current access period. Please do not hesitate to contact me, or

[the] APS Director of Publications, if you have any questions about this important issue.

Recently, the NIH made it mandatory to deposit accepted manuscripts to PubMed Central for all NIH-funded investigators. The APS has maintained its stance that it will accept deposit as long as public access is delayed for 12 months; in fact, APS journals now automatically deposit articles and specify the 12-month delay rather than leaving this decision to authors. As of this writing, however, many journals and societies have not adjusted their copyright statements to account for the NIH policy, and some universities now urge authors to amend, or refuse to sign, journal copyright transfer agreements to avoiding placing themselves in a legally untenable situation.

8.3   PATENTS

In 1970, only three major research universities devoted at least one half-time staff position to technology transfer and research universities as a group secured only ~150 patents (Sampat, 2006). Most universities consciously confined their activities to the generation and free exchange of knowledge; they avoided the business aspects of translating that knowledge into profitable products for fear that it would taint their academic missions. Among others, Columbia University, Harvard University, Johns Hopkins University, The University of Chicago, and Yale University specifically prohibited the patenting of results from biomedical research. At that time, private companies performed most federally funded research and development, and patents derived from that research belonged to the federal government. University owned patents typically resulted from industry funded research and were designed primarily to protect against misuse of the technology. Licensing was assigned via independent foundations or corporations (Sampat, 2006).

Today's landscape for technology transfer differs dramatically. Most research universities operate substantial technology transfer offices, securing and licensing patents covering a range of ideas and products derived largely from federally funded research. A network of consulting, research, and technology transfer agreements links universities to companies, and ownership of intellectual property is frequently the critical concern during the negotiation of such agreements. Professors routinely form "spin-off" companies to translate their discoveries into products. Every practicing scientist and engineer, whether in industry or academia, must now learn the basics of intellectual property law and technology transfer. In fact, if you accept a faculty position at a research university tomorrow, it is likely that the second piece of paper you will sign (after your offer letter) will be a patent agreement.

Exercise 8.3   Economists consider a strong university system to be a major driver of innovation and economic growth. Make a list of the ways that universities transfer information and technology to companies. Next, order your list starting from the mechanism you consider most important. Finally, compare your list to one based on surveys of managers of industrial research and development [see Table 4 in Cohen WM, Nelson RR, Walsh JP (2002) Links and impacts: The influence of public research on industrial R&D. *Management Sci* 48: 1–23]. Discuss with a colleague possible reasons for the main discrepancies between your list and the one compiled by Cohen et al. (2002).

Maintaining our focus on communication in science and engineering, we restrict the remainder of our discussion of patents and technology transfer to two small aspects of this very broad topic: private ownership of patents derived from publicly funded research and the impact of university technology transfer efforts on scientific communication.

8.3.1   Patents and Publicly Funded Research

Before 1980, the federal government owned patent rights to any discovery made with federal funding. The simple rationale for this policy was that inventions generated with public funds should belong to the public. As we discussed for data and models in Section 8.1 and for publications in Section 8.2, however, the general question of ownership is not simple. As with data and publications, generating patents not only requires funding but also knowledge, ingenuity, hard work, equipment, space, and other resources. Hence, faculty, students, the university, and the government might all credibly claim at least partial ownership of a patent based on publicly funded research performed in a university laboratory.

In 1980, the U.S. Congress passed the Bayh–Dole Act, which granted universities and small businesses the rights to patents arising from their federally funded research. From an ownership perspective, this policy was balanced better than the one it replaced — it granted patent rights to universities, required universities to share royalties with the inventors, and retained limited rights for the government. Yet, proponents of the change cited the need to stimulate innovation rather than the need to attribute ownership properly. At that time, federal funding agencies often transferred their patent rights to universities and companies, but each agency had a different policy. The Bayh–Dole Act aimed to replace the array of existing policies with a single, uniform policy. Supporters argued that the government failed to promote licensing and use of the patents it owned, and transferring ownership to companies and universities would promote greater dissemination and utilization of innovations generated from federal funding.

Exercise 8.4   Read "Patenting and U.S. Academic Research in the 20th Century: The World Before and After Bayh–Dole" [Sampat BN (2006) *Res Policy* 35: 772–789], then research the impact of the Bayh–Dole Act. Write a one-page position paper arguing that Bayh–Dole either has or has not enhanced commercialization and utilization of the results of federally funded research in your field.

8.3.2   Patents and Publication

We discuss here a typical process for technology transfer at a research university, but please note that each university and company has its own patent policies. If you have a new idea that you think might merit a patent, you begin by filing an invention report with your university's office of technology transfer. A member of that office will review your report, discuss the idea with you, help research whether existing patents already cover your idea, and make a decision about whether to proceed. If positive, the next step is usually to file a provisional patent application. Filing a provisional patent is relatively inexpensive and protects your idea for 12 months while the university decides whether to proceed with a full patent application. Subsequent actions differ widely depending on the university and the nature of the invention. Because it can be expensive to file a full patent application, some universities spend the 12 months shopping your idea to potential corporate partners and proceed only if a partner is willing to license the patent once awarded. This approach works well when your invention is developed fully, a prototype has been tested, or the new idea has such obvious value that an investor will commit based on the idea alone. University-based research commonly produces ideas at a much earlier stage of development, however. In these cases, it is often difficult to decide whether to invest in a full patent application based on the limited information at hand.

After reviewing your invention report, the office of technology transfer may recommend that you delay initiating the patent process until you have developed and tested your idea further. This approach is where the patent process can begin to conflict with the normal practice of academic research. Because any public disclosure of the idea impacts the patent process, it is critical that you discuss your ideas with your office of technology transfer before submitting them for publication or presentation at professional meetings. Without publications, however, it may be difficult to obtain additional funding to mature your idea, and even a grant application might constitute a public disclosure in some circumstances. There is also the risk that someone else will advance a similar idea while you wait.

No one, including those in your office of technology transfer, can tell you how best to balance such concerns. In general, we recommend entering any discussion of technology transfer with a clear vision of your research and career goals, communicating those goals to your technology transfer officer, and doing your best to make decisions consistent with those goals. If, for example, your ultimate goal is to make a significant impact on the treatment of cancer, patenting a new drug or method of drug delivery and marketing it to companies may be an integral part of achieving your goal. In contrast, if your goal is to design a simple and effective water filter that can be assembled cheaply and easily in the developing world, posting your design on a Web site and publicizing it through the press or nonprofit agencies might be a more effective approach.

Concerns of intellectual property also impact scientific communication when industry funds academic research. Universities negotiate the terms of research contracts and agreements with sponsoring agencies and companies. Large federal agencies such as the NIH can essentially dictate terms to universities, but individual research agreements with companies vary widely. Intellectual property rights are often the major focus of negotiations, and restrictions on publication and other dissemination of the results of the research are common. Companies often demand the right to prereview and block any planned public disclosure, including abstracts, conference presentations, journal publications, and grant applications. Few universities agree to such stringent limits, but many agree to a waiting period to give the sponsor adequate time to review any planned disclosure and file appropriate patent applications. As a principal investigator of an industrial sponsored project, it is critical to work closely with the contract negotiating team at your university to make sure you understand and are willing to accept any proposed limits on publication. As a student or postdoctoral fellow considering whether to work on an industrial sponsored project, it is essential to ask whether the project includes restrictions on publication, for your ability to publish is critical to building your track record and thus your career.

8.4   PLAGIARISM

Most scientists and engineers would tell you that they know what constitutes plagiarism, that they consider it a serious offense, and that they would never do it, suggesting that plagiarism is not a major problem among working scientists and engineers. Many faculty members would admit that there is more of a problem with plagiarism among university students, but they would attribute this primarily to two factors: the Internet, which provides easy access to text written by others, and students from cultures that have different conventions regarding how and when to incorporate or cite ideas from published work. Yet, data from recent surveys contradict these common perceptions. Undergraduates in the United States are frequently confused over what constitutes plagiarism, they do not consider it a serious form of cheating, and they do it with shocking frequency. Although scientists self-report much lower rates of plagiarism, they report frequent observations of plagiarism by colleagues, suggesting that plagiarism is a significant problem, and not just among students.

Studies on undergraduate cheating by McCabe and colleagues, in association with the Center for Academic Integrity, provide an interesting introduction to student attitudes about plagiarism (McCabe et al., 2001; McCabe, 2005). This group conducted a series of surveys of undergraduates at universities within the United States and Canada and reported that more than 75% of students admit to some type of cheating. In particular, one study revealed that 26% of students admitted to committing "plagiarism" in the past year, while twice as many (54%) admitted that they "copied one or two sentences without footnoting." A more recent survey found no evidence that the Internet is to blame: 38% admitted to "paraphrasing/copying a few sentences . . . without footnoting" from a written source, whereas 36% admitted to copying from an Internet source. Apparently many undergraduates simply do not feel that this as a serious offense; only 56% rated plagiarism as moderate or serious cheating (versus trivial or not cheating; McCabe, 2005). It may be surprising that graduate students took plagiarism only slightly more seriously: 32% considered paraphrasing or copying a few sentences without footnoting to be trivial or not cheating and 25% admitting to doing it in the past year. Even more concerning than the high rates of self-reported plagiarism in these studies is that the students surveyed appeared to have a narrow understanding of plagiarism as direct word-for-word copying, believing that it was acceptable to use someone else's ideas without attribution as long as they expressed those ideas in their own words (McCabe et al., 2001).

Whether it is called plagiarism or simply misconduct, "Using another's ideas without obtaining permission or giving due credit" is a pretty good working definition of a key problem that arises often in science and engineering. De Vries et al. (2006) found that although only 1.4% of NIH-funded scientists admitted to using another's ideas within the past 3 years without giving credit, 45.7% reported observing this behavior among their colleagues over the same period.

Exercise 8.5   "Using another's ideas without obtaining permission or giving due credit" covers a wide range of potential behaviors beyond direct word-for-word copying of published text. Perhaps a colleague suggested an interesting experiment during a conversation at a conference, prompting you to perform that experiment and publish the results without further discussion with your colleague. Is this misconduct? What about testing a hypothesis suggested in the discussion section of a paper you read, then publishing your findings without citing the paper that suggested the hypothesis? Together with a colleague, list 10 examples of using another's ideas without permission or credit, then decide which you consider to be appropriate or inappropriate. Compare your list with colleagues to determine the most frequently listed examples. Include within your discussion the concept of common knowledge versus personal intellectual property.

Problems that arise commonly in a discussion of using another's ideas involve interactions with group members, citation, and peer review. We discuss the first two categories briefly below and peer review in the following section.

8.4.1 Attribution Within a Research Group

Many issues regarding attribution within a group are addressed by the discussion of authorship in Chapter 6. Ideally, every group member should receive proper credit on publications through appropriate authorship or acknowledgment. Yet, ideas conceived within a group are often presented in other venues. Must a professor giving an academic seminar explicitly list the names of all students and fellows who gathered the data presented? Is it sufficient to acknowledge the entire group on a slide at the end, as is common, or should the advisor specifically attribute each graph and figure, as with figures taken from published work by other groups? Such questions become particularly delicate as postdoctoral fellows approach the transition to an independent academic career. Fellows may view the use of their ideas in their adviser's grant applications as appropriation, while their adviser may see these ideas as belonging to him or her as principal investigator of the group.

8.4.2   Citation

It is clearly unethical to omit citations of relevant work intentionally, whether to claim undue credit for previously published ideas or to slight the work of a rival. Omission of an important reference is more commonly an honest mistake — the author simply misses an important paper in the exponentially expanding sea of archival literature. That a mistake is honest does not lessen its impact, however. Omission of a key reference deprives deserving colleagues of credit for their ideas and misleads readers interested in the topic. In medical malpractice cases, actions are judged according to the "standard of care," that is, what most physicians would do in a given situation. If most researchers diligently review the relevant literature and find key references before writing an article, should missing an important reference be considered research misconduct?

All of us have experienced another problem related to citation: while reading an article, we encounter an interesting statement referenced to an earlier publication, retrieve the original reference, and find that it says something very different than was claimed. In some cases, this may be an honest difference of opinion; two researchers reading the same article may interpret its key findings differently. Frequently, however, discrepancy between attributed content and actual content arises through a scientific version of the "telephone game" — a chain of citation, where each author depends on a previous citation rather than retrieving and reading the original paper, propagates an error in describing the content of that paper. As with omitted references, inaccurate citation misleads the reader and misrepresents the work of colleagues. Such errors are not considered actionable misconduct by universities or funding agencies, yet their impact on the archival literature and on your reputation can be significant.

Exercise 8.6   High-profile plagiarism cases occur regularly in science and engineering as well as in history, literature, and other fields. Find and evaluate one recent case in science or engineering and one case outside science and engineering. For each case, write a one-page summary, including what the author plagiarized, how the plagiarism was detected, any explanations offered by the author, and the impact of the plagiarism on the author's career.

8.5 PEER REVIEW

Science and engineering rely heavily on peer review, the evaluation of your work by your peers. Peer review is central to deciding which papers a journal will publish, which grants agencies will fund, which patents will be issued, and which drugs and devices will be approved. Any system of peer review must balance the fact that colleagues who work in your field are best qualified to review your work against the possibility that those colleagues include former students or mentors as well as current collaborators or competitors. Evaluating and avoiding conflict of interest is integral to effective peer review.

Exercise 8.7 Imagine you have been named editor of a new journal in your field. Formulate a policy stating how your journal will handle potential conflicts of interest when assigning reviewers. What constitutes a conflict of interest? Are all conflicts equal? How will conflicts be handled? Must any reviewer with a potential conflict decline to review, or are there some situations where declaring the conflict will suffice? Will you hire staff to search for potential conflicts or will you rely on reviewers to disclose conflicts? Will you allow authors to name reviewers they consider to be in conflict? What other measures will you take? Summarize your thoughts in a two-page paper.

8.5.1 Archival Journal Articles

A journal editor who receives a manuscript for review usually begins by reading the abstract and scanning the paper to verify that it is appropriate for the journal. The next step is to assign reviewers. Editors typically identify reviewers from a variety of sources, including personal knowledge of investigators in the field, databases maintained by the journal, literature searches using keywords or title words from the manuscript, and the references cited within the manuscript. Some journals allow authors to suggest reviewers at the time of submission.

Selection of appropriate reviewers helps to ensure a fair and thorough review. Typically, editors try to assign several reviewers who are experts in the field but not associated closely with the authors or one another. Achieving this goal is not as easy as it sounds; it assumes that the editor knows professional relationships among those working in the field — who trained in which laboratories, who has collaborated with whom, who has published with whom. Most of this information can be found with enough research, but such research would be too time-consuming for every submitted manuscript. Hence, journals rely on multiple safeguards against conflict of interest. Securing multiple reviewers provides one such safeguard, for it is less likely that multiple reviewers will have a conflict of interest or the same personal bias for or against a particular author. Some journals divide the task of assigning reviewers among multiple associate editors or members of an editorial board to help ensure that assignments are based on detailed scientific and professional knowledge about subfields within a broader discipline. Many journals ask authors to list those whom they feel have a conflict of interest and will often exclude those reviewers.

Allowing authors to name potential reviewers for exclusion raises many interesting questions. As an author, you may have genuine concerns whether a particular competitor will judge your work more harshly than normal. Yet, are you willing to assert to the editor, often a senior colleague in your field, that this rival is incapable of judging your work fairly? Does having the potential to exclude particular reviewers tempt authors to try to avoid valid criticism that ultimately could help improve the research and the paper? Ultimately, much relies on the judgment of the editor. Ideally, an editor who receives two very positive reviews and one negative, but unfair, review will read the reviews, recognize the lack of substance in the negative comments, and discount the unfair review when making a decision. If a journal receives a large number of submissions and is highly selective, however, an editor may simply average scores from all reviewers without resolving potential discrepancies; in such cases, one negative review may be enough to prevent acceptance. Fortunately, there are many journals in each field, so there are always other opportunities to publish a high-quality paper. Recall from Chapter 3, however, that one should always take advantage of any opportunity to improve a paper via revision.

Assigning reviewers and integrating their feedback can be demanding, but it is usually straightforward from an ethical point of view. In contrast, reviewing an unpublished manuscript frequently raises difficult ethical questions. First, you must decide whether to accept the request to review. Generally, you should not agree to review the work of colleagues from your institution, of former students or mentors, or of current or recent collaborators. What if your former student graduated 20 years ago, however, and you have not collaborated since? What if one of the authors is a former collaborator whom you have not published with or spoken to in 5 years? Few journals provide potential reviewers with specific instructions on exactly what constitutes a conflict of interest. Most rely on the judgment of the reviewers and ask them to err on the side of caution. *Maintaining the integrity of peer review thus requires individuals to avoid not only actual conflicts of interest but also apparent conflicts of interest.* In other words, if the author of a manuscript might believe you are biased, you should not review the paper, even if you are confident you can provide an objective review.

You may find it difficult to evaluate objectively the work of a colleague you dislike; most would agree that this constitutes a bias. A more interesting question is whether you should review a manuscript written by a colleague whose work you consider to be generally of poor quality. On one hand, your responsibility as a reviewer is to help the editor evaluate the quality of work submitted for publication and to ensure the publication of only high quality research; if you know the work of a particular group well, you may be uniquely qualified to explain why a particular manuscript does not merit publication. Yet, if you consider a particular group's work to be poor, it is likely that the authors will consider you misinformed or even biased against them. Fortunately, this is another situation where multiple reviewers provide a safeguard. If you explain carefully why you believe the work is flawed, the editor can balance your evaluation against those of other reviewers in making a final decision.

This discussion raises an issue that is critical to consider, especially by novice reviewers. The primary responsibility of the reviewer is to recommend whether to accept or reject a particular manuscript; in most cases, however, an intermediate step is to recommend potential revisions that would improve the work. By explaining carefully the strengths and weaknesses of a manuscript, you not only help the authors improve the manuscript, you also help advance their research and the field in general. Few things are more frustrating to an author than receiving a review that rates the paper poorly without providing specific criticisms. By contrast, few things are more helpful than a critique that suggests a critical new experiment or an alternative interpretation of your data that you had not considered previously. Again, however, we see that the reviewer is faced with an ethical decision. What if you, the reviewer, realize that the authors have missed the most important experiment or calculation? You could argue that the paper should be rejected because the authors have not proved their hypothesis or provided convincing results, then pursue the correct approach on your own and publish it later. Alternatively, you could provide significant guidance to the authors so that they can pursue the correct approach and publish what would be an important result that they would be congratulated for without any recognition of your anonymous contribution. Who would you ask for advice in this situation?

Exercise 8.8   Imagine that you are a third-year Ph.D. student and your faculty advisor asks you to review a paper that he or she was asked to review by a top journal. If you have never reviewed a paper before and have not yet published a paper, how should you proceed? What information would you expect your advisor to provide? If you complete the review and it is to be submitted to the journal based solely on your evaluation, should the editor be so notified? Should you get the "credit" for the review? How do you think the authors would feel if they disagreed with the conclusions of the review, which were very negative, and they learned that it was conducted by a student?

8.5.2   Grants

De Vries et al. (2006) met with a number of focus groups consisting of researchers and formulated a survey that they eventually conducted and reported in the *Nature* article "Scientists Behaving Badly" (Martinson et al., 2005), which was discussed earlier in this book. In the focus groups, they found researchers to be less worried about frank plagiarism or fabrication of data than about handling the "fuzzier" situations that arise in science and engineering, such as excluding data or properly apportioning credit for ideas and discoveries. We have focused much of our discussion in this chapter on these gray areas, where each of us must rely on our values and judgment in the absence of universally accepted rules. Regarding peer review, the primary concern of the researchers interviewed by De Vries and colleagues was potential theft of their ideas during grant review. One commented (De Vries et al., 2006).

> I'm always wary of submitting grants to [NIH] study sections, because those people who sit on the study sections, it's not unknown for them to take your ideas, kill your grant, and then take and do it. And I think all of us have either had that happen to them or know somebody who had that happen to them.

Determining whether to excuse yourself from reviewing a particular grant is often simpler than for a particular journal article because funding agencies usually issue more specific guidelines than journals. At the NIH, for example, you may not participate in the review of any grant application from your institution; if you are on the panel, you must leave the room during the discussion of these applications. You must also excuse yourself from the review of applications involving any colleague with whom you have published during the past 3 years. Furthermore, the review panel on which you sit may not review any application involving you; these grants must be sent to a different panel for consideration. NIH reviewers and staff also work to identify and avoid other conflicts, whether real or apparent, not governed by specific guidelines.

The more challenging issue with regard to reviewing grants is deciding how to handle information contained within the applications. Although each application is confidential and must be destroyed following review, you will remember many of the things that you read (and heard during a panel review). Indeed, in addition to the importance of fulfilling a professional responsibility, reviewing grant applications often benefits a reviewer in three ways: you see the difference between well-prepared and poorly prepared applications, which can help you prepare more competitive applications, you learn more about the overall process and what other reviewers value, and you are exposed to things that you may not have been aware of, including important references, new instrumentation or materials, useful experimental methods or computational tools, and so forth. Indeed, you may even learn of individuals who would be good potential collaborators. Yet, because applications are confidential, what information can you use and when? A good rule of thumb is that *any information available in the public domain can be used in good conscience*, including published papers, commercially available instruments, materials, software, and contacts listed on the Web for individuals in academia and industry. In other words, if you can obtain the same information elsewhere, it can be used.

In contrast, novel ideas (e.g., new experimental protocols or methods to solve a complex equation) that are not available in the public domain should not be used. Some might ask in this regard if it would be acceptable to contact the investigators and ask for permission to use their ideas.

Albeit perhaps surprising, the answer is that you are not supposed to discuss any aspect of the grant application or its review with the applicant(s), hence it is not appropriate to seek such permissions. Rather, one should wait for the ideas to be made public by the investigators, often via a conference presentation, published paper, or posting on the Web. Once such disclosure has been made, it is then acceptable to use the available information or to contact the investigators for further clarification or possible collaboration. If the disclosure appears via a patent rather than via free information, however, one must then respect the conditions of the patent.

That it is appropriate to use information in the public domain should be obvious, but this does not resolve all potential issues. What if the applicants did not get funded and they were unable to pursue their ideas? What if they were funded but decided to pursue a different approach? In other words, is it good for science and engineering in general to let an excellent idea die simply because those who conceived the idea could not bring it to fruition? Is there a statute of limitations on such ideas or should there be? What if you wish to apply their idea to a completely different problem, one they are most likely not interested in and would never pursue? What if you were already planning to do the same experiment or one related closely — do you need to forego your experiment to avoid what may appear to be misconduct? Who should you approach to find answers to such questions?

Exercise 8.9   We identified some situations regarding "intellectual property" that may arise when a person reviews, in confidence, a grant application. Another situation could arise if you submitted a grant application to the NIH that needed to be revised two times (which generally means a two or more year delay in starting the research). What could you do if you suspected that someone on the review panel was purposely trying to delay your research so that they could pursue the same idea? List five other potential situations that could arise in grant reviews and discuss potential ethical issues.

In conclusion, recall from the Preface that it was not the goal of this book to be a standalone source on matters of style or ethics in communication. Rather, our goal was to motivate the reader to develop an effective, individual style of communicating and a personal commitment to integrity because it matters. We sought to raise questions, not answer them. Our best advice is simply to seek advice from good role models whose writing and ethics you respect — learn from others, solicit constructive feedback on oral presentations as well as drafts of manuscripts and proposals, and discuss potential concerns on ethical matters with peers, advisors, supervisors, and administrators. Work hard to ensure that, when you look back over your career upon retirement, you are proud of a job well done.

References

Bell ET (1986) *Men of Mathematics.* Simon & Schuster, New York, NY.

Berry TE (1971) *The Most Common Mistakes in English Usage.* McGraw-Hill, New York, NY.

Bhopal R, Rankin J, McColl E, Thomas L, Kaner E, Stacy R, Pearson P, Vernon B, Rodgers H (1997) The vexed question of authorship: Views of researchers in a British medical faculty. *BMJ* 314: 1009–1112.

Blake G, Bly RW (1993) *The Elements of Technical Writing.* Macmillan, New York, NY.

Boorstin DJ (1983) *The Discoverers.* Vintage Books, New York, NY.

Brogan JA (1973) *Clear Technical Writing.* McGraw-Hill, New York, NY.

Carnegie D (1956) *How to Develop Self-Confidence and Influence People by Public Speaking.* Simon & Schuster, New York, NY.

Clendening L (1960) *Source Book of Medical History.* Dover Publications, New York, NY.

Cohen WM, Nelson RR, Walsh JP (2002) Links and impacts: The influence of public research on industrial R&D. *Management Sci* 48: 1–23. doi:10.1287/mnsc.48.1.1.14273

Davidoff F (2000) For the CSE Task Force on Authorship. Who's the author? Problems with biomedical authorship, and some possible solutions. *Sci Ed* 23: 111–119.

Day RA, Gastel B (2006) *How to Write and Publish a Scientific Paper.* 6th Edition. Greenwood Press, Westport, CT.

De Vries R, Anderson MS, Martinson BC (2006) Normal misbehavior: Scientists talk about the ethics of research. *J Empir Res Hum Res Ethics* 1: 43–50. doi:10.1525/jer.2006.1.1.43

Dillenberger J (1961) *Protestant Thought and Natural Science.* University of Notre Dame Press, Notre Dame, IN.

Eastwood S, Derish P, Leash E, Ordway S (1996) Ethical issues in biomedical research: perceptions and practices of postdoctoral research fellows responding to a survey. *Sci Eng Ethics* 2: 89–114. doi:10.1007/BF02639320

Engler RL, Covell JW, Friedman PJ, Kitcher PS, Peters RM (1987) Misrepresentation and responsibility in medical research. *N Engl J Med* 317: 1383–1389.

Flanagin A, Carey LA, Fontanarosa PB, Phillips SG, Pace BP, Lundberg GD, Rennie D (1998) Prevalence of articles with honorary authors and ghost authors in peer-reviewed medical journals. *JAMA* 280: 222–224. doi:10.1001/jama.280.3.222

Frankel MS (1999) Public access to data. *Science* 283: 1114. doi:10.1126/science.283.5405.1114

Gibaldi J (1995) *MLA Handbook for Writers of Research Papers.* 4th Edition. The Modern Language Association of America, New York, NY.

Iverson C, et al. (1998) *AMA Manual of Style: A Guide for Authors and Editors.* 9th Edition. Lippincott Williams & Wilkins, Philadelphia, PA.

Kilpatrick JJ (1984) *The Writer's Art.* Andrews and McMeel, Kansas City, MO.

Lightman A (2005) *The Discoveries: Great Breakthroughs in 20th Century Science.* Vintage Books, New York, NY.

Martinson BC, Anderson MS, de Vries R (2005) Scientists behaving badly. *Nature* 435: 737–738. doi:10.1038/435737a

Mason SF (1962) *A History of the Sciences.* Collier Books, New York, NY.

McCabe D (2005) Cheating among college and university students: A North American perspective. *Int J Educ Integrity* 1(1) [Epub].

McCabe DL, Trevino LK, Butterfield KD (2001) Cheating in academic institutions: A decade of research. *Ethics Behav* 11: 219–232. doi:10.1207/S15327019EB1103_2

Motz L, Weaver JH (1989) *The Story of Physics.* Avon Books, New York, NY.

Parker RA, Berman NG (1998) Criteria for authorship for statisticians in medical papers. *Stat Med* 17: 2289–2299. doi:10.1002/(SICI)1097-0258(19981030)17:20<2289::AID-SIM931>3.0.CO;2-L

Piwowar HA, Day RS, Fridsma DB (2007) Sharing detailed research data is associated with increased citation rate. *PLoS ONE* 2: e308. doi:10.1371/journal.pone.0000308

Sampat BN (2006) Patenting and U.S. academic research in the 20th century: The world before and after Bayh–Dole. *Res Policy* 35: 772–789.

Shamos MH (1959) *Great Experiments in Physics: Firsthand Accounts from Galileo to Einstein.* Dover Publications, New York, NY.

Strunk W, White EB (1979) *The Elements of Style.* 3rd Edition. Allyn and Bacon, Boston, MA.

Tarnas R (1991) *The Passion of the Western Mind.* Ballantine Books, New York, NY.

Tarnow E (1999) The authorship list in science: Junior physicists' perceptions of who appears and why. *Sci Eng Ethics* 5: 73–88. doi:10.1007/s11948-999-0061-2

Valiela I (2001) *Doing Science: Design, Analysis, and Communication of Scientific Research.* Oxford University Press, New York, NY.

Van Doren C (1991) *A History of Knowledge.* Ballantine Books, New York, NY.

Vivian CH, Jackson BM (1961) *English Composition.* Barnes & Noble, New York, NY.

Walters R, Kern TH (1991) How to eschew weasel words. *Johns Hopkins Magazine,* December 1991.

Westfall RS (1993) *The Life of Isaac Newton.* Cambridge University Press, Cambridge, UK.

Yank V, Rennie D (1999) Disclosure of researcher contributions: A study of original research articles in *The Lancet. Ann Intern Med* 130: 661–670.

Author Biography

Jay D. Humphrey is Regents Professor and Carolyn S. and Tommie E. Lohman Professor of biomedical engineering at Texas A&M University. He has authored a graduate textbook (*Cardiovascular Solid Mechanics*), coauthored an undergraduate textbook with a former student (*An Introduction to Biomechanics*), published more than 150 archival papers and chapters in other books and encyclopedias, and serves as coeditor in chief for the international journal *Biomechanics and Modeling in Mechanobiology*. He has served as a reviewer for 50 technical journals and 20 funding agencies in the United States and abroad. He is a fellow of the American Institute for Medical and Biological Engineering.

Jeffrey W. Holmes is Associate Professor of biomedical engineering and medicine at the University of Virginia. He has published more than 40 archival papers and book chapters and has reviewed for 15 technical journals and several funding agencies, including the American Heart Association and the National Institutes of Health. Before moving to Virginia, he developed and taught the course "Ethics for Biomedical Engineers" at Columbia University, where he won the Distinguished Faculty Teaching Award. Other awards include an Alexander von Humboldt Research Fellowship, the Y.C. Fung Young Investigator Award, and an Established Investigator Award from the American Heart Association.

Index

Printed in the United States
by Baker & Taylor Publisher Services